SPEED
MATHEMATICS

Rajesh Kumar Thakur is a mathematics teacher by profession and also a popular mathematics writer. He has been the Honorary Secretary and Chairman of the Award Selection Committee of All India Ramanujan Maths Club (AIRMC) since 2012. Widely published with more than 56 books, 62 e-Books, 500 articles related to mathematics and 300 blogs to his credit, Thakur is a regular column writer with over 500 published articles in various well-known newspapers and magazines including *Amar Ujala* and *Navbharat Times*.

A number lover, Thakur is the recipient of several awards including the National Best Teacher Award (2010), Arvind Pandey Yuva Lekhan Award (2015), the Math Genius Award (2017) and the Effective National Teacher Award (2018).

He lives in Delhi.

Also by the author

Mathematics in Religion
Maths Made Easy
The Essentials of Vedic Mathematics

SPEED MATHEMATICS

Do It Quick, Do It Right!

Rajesh Kumar Thakur

RUPA

Published by
Rupa Publications India Pvt. Ltd. 2018
7/16, Ansari Road, Daryaganj
New Delhi 110002

Sales centres:
Bengaluru Chennai
Hyderabad Jaipur Kathmandu
Kolkata Mumbai Prayagraj

Copyright © Rajesh Kumar Thakur 2018

While every effort has been made to verify the
authenticity of the information contained in this book,
the publisher and the author are in no way liable for the use of the
information contained in this book.

All rights reserved.
No part of this publication may be reproduced, transmitted,
or stored in a retrieval system, in any form or by any means,
electronic, mechanical, photocopying, recording or otherwise,
without the prior permission of the publisher.

P-ISBN: 978-93-5304-089-5
E-ISBN: 978-93-5304-090-1

Tenth impression 2025

15 14 13 12 11 10

The moral right of the author has been asserted.

Printed in India

This book is sold subject to the condition that it shall not, by
way of trade or otherwise, be lent, resold, hired out, or otherwise circulated,
without the publisher's prior consent, in any form of binding or cover
other than that in which it is published.

Contents

Foreword — *vii*

I Addition & Subtraction: Here's to Effective Modus Operandi
1. Addition — 3
2. Subtraction — 13
3. About Time — 20

II Multiplication, But Without Tears
1. It Is No Puzzle — 27
2. Multiplication Techniques from Around the World — 47
3. Multiplication Using Napier's Rods — 65

III Division, Did You Say 'Insoluble'?
1. The Egyptian Method of Division — 73
2. Auxiliary Fraction — 90
3. Division of Polynomials — 105

IV Fractions
1. As Easy As Pie — 113

V Golden Rules to Your 'Speedy' Rescue
1. Casting Out Elevens vs Casting Out Nines — 135
2. Rule of 72 — 146
3. Finding Prime Numbers — 150
4. Finding the Unit Digit — 154

5.	The Magic of the Pascal Triangle	159
6.	Finding the Area of a Polygon	167

VI Squares and Cubes are Much Fun

1.	Square	183
2.	Square Roots	204
3.	Cube	218
4.	Cube Roots	230

Foreword

We know from our experience and research that children are more likely to be successful learners if they enjoy the learning process. The world is increasingly becoming technology-oriented and to have to grow up in such a scenario, it is even more important that students have strong mathematical skills not only to help them in their work life, but also in their everyday life.

To hone their math skills, one must be able to appreciate and enjoy different methods to solve problems. This requires that the methods be presented in a way that is appealing and easy. Classroom textbooks are filled with standard algorithms with little or no variation in their presentation. With technology entering the classrooms, concepts are now being presented attractively, but to develop fluency in applying them, approaches other than the standard algorithms, need to be mastered.

It is in this context that Shri Rajesh Thakur has researched and put together a variety of easy math calculation methods in this superb book *Speed Mathematics*. And that makes it a 'must have' for students.

Presenting these methods within a historical context also makes it an interesting must-read for both the young and old. The numerous examples that have been worked out to show how the various speed calculation methods can be applied are sure to influence the reader and better his/her repertoire of mathematical skills.

Read the book, try the new methods and enjoy mathematics like never before!

MEENA SURESH
Director
Ramanujan Musuem and
Maths Education Centre, Chennai

I
ADDITION & SUBTRACTION
HERE'S TO EFFECTIVE MODUS OPERANDI

Addition

Addition is the most fundamental operation carried out in mathematics by everyone. Be it a daily wage worker or an industrialist, everyone has to perform this operation in his life almost every day. Though there are no shortcut methods that can tell you the result instantly, there are certain methods that can enhance your ability to calculate faster and more accurately.

The basic problem we face in the process is that we are not good at crunching numbers. We learn the multiplication table to speed up the multiplication process but I wonder why 99 per cent of children in our schools aren't aware of the addition table. The knowledge of the latter can make a big difference when it comes to doing additions, since, like the multiplication table, addition tables act as ready reference with the help of which you can add two or three single digits.

So let's start with the different methods of adding numbers.

Now, we are accustomed to add from right to left, i.e. we start adding the first column from the right and then move backwards, but have you ever tried to add from the right or from the middle? Let me take a few examples to help you see the modus operandi behind such a process.

Example: Add 78 and 45.
Solution: Start adding from the left and add the numbers in the first column from the left. Here, in the first column, we have to add 7 and 4 first.

```
  7 8
+ 4 5
-----
  11
```

Next, add the digits 8 and 5 in the ones' place. Write the sum: 8 + 5 = 13 in the second column as shown here, with a comma or any separator in between.

```
    7  8
+   4  5
---------
   11 / 13
```

Finally, add the middle digits as shown here. Retain the extreme digits as they are.

```
   7  8
+  4  5
--------
  11, 13
    ⌣
    +
  = 123
```

Example: Add 17 + 18.
Solution: Write the numbers one below the other. Add the numbers in the left column and then add the ones in the right. Finally, add the middle digits.

```
  | 1 | 7 |        | 1 | 7 |
+ | 1 | 8 |      + | 1 | 8 |
-----------      -----------
  | 02 , 15 |      02 , 15
                      ⌣
                      +
                   = 35
```

Example: Add 365, 472 and 232.
Solution: Write each number one below the other and add column wise. Once the column wise sum is done, you can add the middle digits to get the final answer.

3	6	5
4	7	2
+2	3	2
9,	16,	9

Remember one thing, while adding the numbers of each column, every column wise sum should have two digits each. In case the sum of the digits is less than two digits, add a zero before the sum.

3	6	5
4	7	2
+2	3	2
09,	16,	09

Now, add the middle digits.

3	6	5
4	7	2
+2	3	2
09,	16,	09

$$= 09, 16, 09$$
$$= (9+1)/(6+0)/9$$
$$= 1069$$

Example: Add 73727 + 32541 + 41324 + 62515 + 20804

Solution: First write the numbers one below the other. Add them column wise.

7	3	7	2	7
3	2	5	4	1
4	1	3	2	4
6	2	5	1	5
2	0	8	0	4
22	08	28	09	21

Now you can add the tens' digit from each column to the digit in the preceding column to get the final sum.

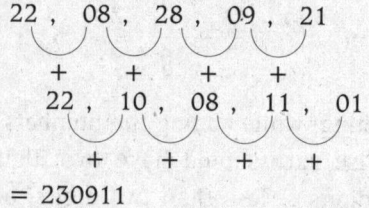

= 230911

Example: Add 38976 + 67907.
Solution: Write the numbers one below the other and add each column separately.

	3	8	9	7	6
+	6	7	9	0	7
	9	15	18	07	13

Add the digits in the tens' place of each column to the digit in the preceding column to get the final sum.
9 / 15 / 18 / 07 / 13

$$9 / 15 / 18 / 07 / 13$$
$$= 9 + 1 / 5 + 1 / 8 + 0 / 7 + 1 / 3$$
$$= 106883$$

Let's explore the addition method discussed with some changes.

The sum of each column is first derived separately. These sums are then placed one under the other with each successive sum being set one space to the left.

Example: Add 38976 + 67907.
Solution:

	3	8	9	7	6
+	6	7	9	0	7
	9	15	18	07	13

Now, the sum of each column is placed one below another and each successive sum set one space to the left:

$$\begin{array}{r}13\\07\\18\\15\\+09\\\hline 106883\end{array}$$

Example: Find the sum of 4598 + 679 + 4858 + 3267.
Solution:

$$\begin{array}{r}4\;5\;9\;8\\6\;3\;7\;9\\4\;8\;5\;8\\+\;3\;2\;6\;7\\\hline\end{array}\qquad\begin{array}{r}4\;5\;9\;8\\6\;3\;7\;9\\4\;8\;5\;8\\+\;3\;2\;6\;7\\\hline 17\;18\;27\;32\end{array}\qquad\begin{array}{r}32\\27\\18\\+17\\\hline 19102\end{array}$$

Example: Find the sum of 628764 + 876938 + 549087.
Solution:

$$\begin{array}{r}6\;2\;8\;7\;6\;4\\8\;7\;6\;9\;3\;8\\+\;5\;4\;9\;0\;8\;7\\\hline 19\;13\;23\;16\;17\;19\end{array}\quad\Longrightarrow\quad\begin{array}{r}19\\17\\16\\23\\13\\+19\\\hline 2054789\end{array}$$

OR

$$\begin{array}{r}6\;2\;8\;7\;6\;4\\8\;7\;6\;9\;3\;8\\+\;5\;4\;9\;0\;8\;7\\\hline 19\;13\;23\;16\;17\;19\end{array}$$

$$= 2\;0\;5\;4\;7\;8\;9$$

Example: Find the sum of 7654 + 9278 + 6949 + 3827 + 1403.
Solution:

```
    7  6  5  4                    3 1
    9  2  7  8                    1 8
    6  9  4  9      ⟹             2 9
    3  8  2  7                   +2 6
   +1  4  0  3                  ───────
   ───────────                   2 9 1 1 1
   26 29 18 31
     (+)(+)(+)
  = 2 9 1 1 1
```

Hope you have enjoyed this process of doing additions.

The whole operation can be done quickly if you learn the addition table; that way, most of your problems in addition will get solved. Try to remember the following table. A 10 × 10 addition table has 100 entries with only 20 numbers repeated throughout the table. Remembering the addition table will make the calculation process easy as you need not count the sum of digits each time on your fingers. Thus it doubles your calculation speed.

ADDITION TABLE										
+	**1**	**2**	**3**	**4**	**5**	**6**	**7**	**8**	**9**	**10**
1	2	3	4	5	6	7	8	9	10	11
2	3	4	5	6	7	8	9	10	11	12
3	4	5	6	7	8	9	10	11	12	13
4	5	6	7	8	9	10	11	12	13	14
5	6	7	8	9	10	11	12	13	14	15
6	7	8	9	10	11	12	13	14	15	16
7	8	9	10	11	12	13	14	15	116	17
8	9	10	11	12	13	14	15	16	17	18
9	10	11	12	13	14	15	16	17	18	19
10	11	12	13	14	15	16	17	18	19	20

The Double Column Method

So far we have discussed the single column method where we were dealing with each column separately but this can be made easier with the double column method. In this method, we break up the numbers to be added in columns taking two numbers in each column. It will be more useful if you can break the numbers in multiples of 10.

Example: Add 16 + 49 + 194 + 66 + 21 + 24.
Solution
Step 1: Here, the complete observation shows us that 16 + 24 is likely to yield a rounded result, i.e. the unit digit is 0. The same would be the case if 49 and 21 were paired. Moreover, the pair 194 and 66 yields a rounded result.

$$16 + 49 + 194 + 66 + 21 + 24$$

Step 2: Rearrange the pair and add as per the pairing done above.

= (16 + 24) + (49 + 21) + (194 + 66)
= 40 + 70 + 260

= (40 + 260) + 70
= 300 + 70
= 370

Example: Add 216 + 159 + 94 + 36 + 41 + 54.
Solution
Step 1: Here, the complete observation shows us that 216 + 94 is likely to yield a rounded result. The same would be the case if 159 and 41 were paired. Moreover, the pair 94 and 26 yields a unit digit of 0.

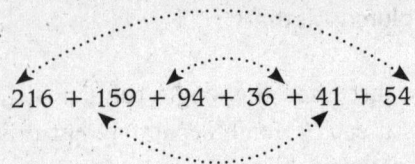

Step 2: Rearrange the pair and add as per the pairing done above.

= (216 + 54) + (159 + 41) + (94 + 36)
= 270 + 200 + 130

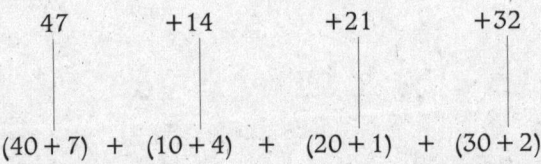

= (270 + 130) + 200
= 400 + 200
= 600

Example: Add 47 + 14 + 21 + 32.

Solution: First break each number into multiples of 10 and a unit.

    47   +14   +21   +32

(40 + 7) + (10 + 4) + (20 + 1) + (30 + 2)

Now apply the commutative property of addition and get the answer.

= (40 + 10 + 20 + 30) + (7 + 4 + 1 + 2)
= 100 + 14
= 114

Suppose you have to add 2345 + 5768 + 8764 + 9127, we will break these numbers into two parts and keep adding two numbers by breaking each of them in multiples of 10 so that we can do the addition conveniently.

$$
\begin{array}{l}
2+23+50+7=82 \\
82+87=80+80+9=169 \\
169+91=169+1+90=260
\end{array}
\quad
\begin{array}{r|r}
 & 2 \\
23 & 45 \\
57 & 68 \\
87 & 64 \\
+91 & 27 \\
\hline
260 & 04
\end{array}
\quad
\begin{array}{l}
45+68=40+60+5+8=113 \\
113+64=110+60+7=177 \\
177+20+3+4=204
\end{array}
$$

Let's take another example.

Add 1159 + 4624 + 6576 + 9783.

Solution:

$$
\begin{array}{l}
2+11+46=40+10+9=59 \\
59+65=50+60+9+5=124 \\
124+97=124+100-3=221
\end{array}
\quad
\begin{array}{r|r}
 & 2 \\
11 & 59 \\
46 & 24 \\
65 & 76 \\
97 & 83 \\
\hline
221 & 42
\end{array}
\quad
\begin{array}{l}
59+24=50+20+9+4=83 \\
83+76=80+70+3+6=159 \\
159+83+9+150+50+30+3=242
\end{array}
$$

This, of course, is not all. You can use a variety of other methods to do addition.* Keep exploring mathematics with innovative ideas!

Practice Problems

Add the following by using the appropriate method:

1. 75934 + 87628 + 34879 + 14093 + 256
2. 876549762 + 345982769 + 470154897 + 284579657 + 145469885
3. 762874 + 3476928 + 593487 + 8752546 + 274039
4. 4876 + 8752684 + 187049 + 48998
5. 6938789159689 + 5248792300000 + 7895248301554 + 8736200145932 + 5260148530489

---

*You can explore some other methods in my previously published titles such as *The Essentials of Vedic Mathematics* (Rupa Publications, 2013) and *Maths Made Easy* (Rupa Publications, 2015).

6. | Km | m |
    |---|---|
    | 256 | 145 |
    | 253 | 874 |
    | 326 | 099 |
    | + 2 | 450 |

7. | Kg | g |
    |---|---|
    | 87 | 958 |
    | +11 | 025 |
    | +05 | 652 |
    | + 1 | 006 |

8. | ₹ | P |
    |---|---|
    | 325 | 25 |
    | +625 | 32 |

9. | ₹ | P |
    |---|---|
    | 659 | 38 |
    | +968 | 15 |

Subtraction

Subtraction is the second fundamental operation in mathematics. It uses a minus sign (–). The word 'subtraction' is derived from the Latin word 'Subtrahere' which is a compound word comprising 'sub' and 'trahere'. 'Sub' means 'under' and 'trahere' means 'to pull down' or 'to take away'.

Of the methods I will be discussing in this chapter,* first let me begin with the Austrian Method.

Austrian Method of Subtraction

The basic problem with subtraction is borrowing. In borrowing, we need to regroup the numbers by decreasing the preceding digit by 1, and the overall look of a subtraction problem appears somewhat to be like—

$$
\begin{array}{rrrr}
 & 14 & 13 & \\
1 & \cancel{4} & \cancel{3} & 11 \\
\cancel{2} & \cancel{5} & \cancel{4} & \cancel{1} \\
-1 & 6 & 7 & 8 \\
\hline
 & 8 & 6 & 3 \\
\end{array}
$$

---

*In one of my earlier books titled *The Essentials of Vedic Mathematics* (Rupa Publications, 2013), I have written in detail about a few Vedic methods that you can check. In another book of mine titled *Maths Made Easy* (Rupa Publications, 2015), I give a detailed description about some interesting methods for subtraction, but here, I shall discuss other methods not discussed in the two books mentioned above.

General Method	Austrian Method

```
        3  12                           4   ₁2
        4̶   2̶                          -²1̶   8
       -1   8                          ─────
       ─────                            2   4
        2   4
```

The Austrian method is neat and easy to understand. How does it work? The basic difference between the general method of subtraction being taught in schools and the Austrian method is that here, instead of decreasing the digit in the next place on the top by 1, we increase the digit in the next place on the bottom by 1.

Let's take another example in detail.

Example: Subtract 247 from 561.
Solution: 561 − 247
   = 500 + 60 + 1 − (200 + 40 + 7)
   = (500 + 60 + **10** + 1) − (200 + 40 + **10** + 7) {10 is added in both groups}
   = (500 + 60 + 11) − (200 + 50 + 7)
   = (500 − 200) + (60 − 50) + (11 − 7)
   = 300 + 10 + 4
   = 314

Let's take one more example.

Example: Subtract 1874 from 4003.
Solution: 4003 − 1874
   = (4000 + 3) − (1000 + 800 + 70 + 4)
   = (4000 + 1000 + 100 + 10 + 3) − (1000 + 1000 + 800 + 100 + 70 + 10 + 4)
   = (4000 + 1000 + 100 + 13) − (2000 + 900 + 80 + 4)
   = (4000 − 2000) + (1000 − 900) + (100 − 80) + (13 − 4)
   = 2000 + 100 + 20 + 9
   = 2129

As you can see, subtraction in the above example has been done smoothly by adding a number of choice on both sides. While actually engaging in the process of subtracting, you need not write out so many steps as explained above; this is only for the purpose of understanding. Let's see the working of Austrian Method in short.

Example: Subtract 2468 from 4743.
Solution:
$$\begin{array}{r} 4\,7\,4\,3 \\ -2\,4\,6\,8 \\ \hline \end{array}$$

Since in the very first column 3 < 8, we can't subtract. We need to simply carry it over. Here, we shall put 1 before 3 in the first column and place 1 at the top of 6 in the second column. It will make subtraction of 8 from 13 easy. Write 13 − 8 = 5 in the first column.

$$\begin{array}{cccc} 4 & 7 & 4 & {}^1 3 \\ -2 & 4 & 6^1 & 8 \\ \hline & & & 5 \end{array}$$

Now, 1 above 6 in the second column has made its value as 6 + 1 = 7. Since in the second column we have 4 − 7, we have to regroup the numbers in the second column. Put 1 before 4 in the second column, making it 14, and place 1 in the third column above 4. Subtracting 7 from 14 in the second column is easy; put 14 − 6 = 8 in the second column. In the third column, 7 − 5 = 2 and in the fourth column 4 − 2 = 2 is placed, making the subtraction process complete.

$$\begin{array}{cccc} 4 & 7 & {}^1 4 & {}^1 3 \\ -2 & {}^1 4 & 6^1 & 8 \\ \hline 2 & 2 & 7 & 5 \end{array}$$

Example: 1953 − 1491 = ?
Solution: In one line, the whole process is done here.

$$\begin{array}{cccc} 1 & 9 & {}^15 & 3 \\ -1 & {}^14 & 9^1 & 1 \\ \hline \times & 4 & 6 & 2 \end{array}$$

Explanation: In the first column: − 3 − 1 = 2

In the second column, 5 < 9, so 1 is placed at the top of 5 and, moreover, 1 is also placed above 4 in the third column.

Second Column: 15 − 9 = 6

Third column: 9 − 5 = 4

Fourth column: 1 − 1 = 0

Hence, 1953 − 1491 = 462

Example: 6052 − 4568 = ?
Solution: The whole process can be done in a single line with three different steps. In the first column, 2 < 8, so 1 is placed before 2 making it 12, and in the second column, the number at the bottom 6 is increased by 1.

Again, in the second column, 5 < 6 + 1 = 7, so we need to regroup the numbers one more time and hence 1 is placed before 1 in the second column, and in the third column before 5. Similarly, in the third column, 6 can't be subtracted from 0; hence 1 is placed before 0 in the third column and 1 is placed over 4 in the fourth column. All this grouping will help avoid borrowing and make the subtraction process easy.

$$\begin{array}{c|c|c}
\begin{array}{cccc} & 2 < 8 & & \\ 6 & 0 & 5 & {}^12 \\ -4 & 5 & {}^16 & 8 \\ \hline & & & 4 \end{array}
&
\begin{array}{cccc} & & 5 < 7 & \\ 6 & 0 & {}^15 & {}^12 \\ -4 & {}^15 & {}^16 & 8 \\ \hline & & 8 & 4 \end{array}
&
\begin{array}{cccc} & & 0 < 6 & \\ 6 & {}^10 & {}^15 & {}^12 \\ -{}^14 & {}^15 & {}^16 & 8 \\ \hline 1 & 4 & 8 & 4 \end{array}
\end{array}$$

Complementary Method of Subtraction

In my previous book, *The Essentials of Vedic Mathematics*, I had discussed a complementary method which was based on a complement of a number from 10.

(1, 9), (2, 8), (3, 7), (4, 6), (5, 5) are complementary as $1 + 9 = 2 + 8 = 3 + 7 = 4 + 6 = 5 + 5 = 10$.

The complementary method of subtraction discussed here is used in Europe and is not a Vedic method. In this, we add some power of 10 (10, 100, 1000, 10000...) in the middle of the problem and then we subtract it out at the end of the problem. The subtraction process begins with the middle number. We subtract the original number at the bottom from the middle number and add the result to the top number. If there is any carrying left out in the whole process, we add the carried number to the top number. Repeat the process until the subtraction process is complete. The leftmost digit of the answer will always be a 1. Cross out the leftmost 1 while writing the final answer.

Example: Subtract 7 from 12.

Solution: Here 2 < 7, so 10 is added in the middle. 0 is placed in the tens' column and 10 in the unit column. Subtract 7 from 10 and add the result to the top number. Write the final result at the bottom.

$10 - 7 + 2 = 5$. Again $0 - 0 + 1 = 1$ is written in the final answer column. Since our subtraction process is complete, 1 in the leftmost column is crossed out.

	1	2		1	2
−		7		0	¹0
			−		7
				~~1~~	5

Hence, $12 - 7 = 5$

Example: Subtract 29 from 47.

Solution: 7 < 9, so we have to add a number in the middle. Here, 100 = 9 tens + 10 ones is added in the middle. As discussed above, 10 − 9 = 1 is added to the number 7 at the top, making the final result of the first column 1 + 7 = 8. In the second column, 9 − 2 = 7 is added to 4 at the top and 11 is written in the answer column. The leftmost 1 is cancelled out, making the final answer 47 − 29 = 18.

```
   4   7              4    7
 − 2   9              9   ¹0      90 + 10 = 100
 ───────            − 2    9
                    ───────
                    ̶1 1    8
```

Example: Subtract 1349 from 2467.

Solution: To solve this problem, 10000 is added in the middle. In the first column, 10 − 9 is added to 7 at the top, making the answer of the first column 8. Similarly, in the second column, 9 − 4 = 5 is added to 6 at the top, making the result 11. 1 of 11 is written in the answer column and 1 is placed at the top of the third column. In the third column, 9 − 3 + 4 + 1 (carry over) = 11; 1 is written in the answer column and 1 (carry over) is placed at the top of the fourth column. In the fourth column, 9 − 1 + 2 + 1 (carry over) = 11 is written in the answer column. The leftmost 1 is cancelled out, making the subtraction process complete.

```
  2   4   6   7        ¹2  ¹4   6   7
− 1   3   4   9         9   9   9  10      9990 + 10 = 10000
───────────────       − 1   3   4   9
                      ───────────────
                       ̶1   1   1   8
```

Hence, 2467 − 1349 = 1148

Example: 56789 − 26987 = ?
Solution:

```
  5 6 7 8 9        5 6 ¹7 ¹8 9
 -2 6 9 8 7        9 9  9  9 10
                 - 2 6  9  9  7
                 ̶1 2 9  8  0  2
```

99990 + 10 = 100000

The last problem is self-explanatory. In the middle, 99990 + 10 = 100000 is added and the subtraction and addition are done as explained above. In the first column, 10 − 7 + 9 = 12; so 2 is written in the answer column and 1 (carry over) is placed at the top. In the second column, 9 − 9 + 8 + 1 (carry over) = 10. The tens' digit 1 is placed at the top of the third column and 0 is placed in the answer part of the second column. In the third column, 9 − 9 + 7 + 1 (carry over) = 8 is written in the answer column. In the fourth column, 9 − 6 + 6 = 9 is placed in the answer column. In the fifth column, 9 − 2 + 5 = 12 is written in the answer column. As directed, the leftmost 1 is cancelled out.

Hence, 56789 − 26987 = 29802

Practice Problems

a) 12345 − 4567
b) 86406 − 37606
c) 9000045 − 7865745
d) 88888888 − 24569997
e) 5674098 − 2487699
f) 784986 − 470098
g) 6000000 − 459762
h) 49020948201 − 25684210976
i) 58747 − 25619
j) 765789 − 467899
k) 7689021 − 2998761
l) 875444 − 768762
m) 12476 − 5698
n) 672199 − 248767
o) 178769 − 98256
p) 20789 − (89214 − 78120)
q) 7612 − (2459 − 1289)

About Time

Adding or subtracting of 'time' is not as easy as adding simple numbers. That is because 60 minutes equals 1 hour. Let's begin with adding numbers in this genre and look at an example related to it.

Adding Time

Add 4 hours 45 minutes and 7 hours 57 minutes

Solution:

	Hrs	Mins
	4	45
	+7	57
	11	102
=	11+1	102−60
=	12 hrs	42 min

You can see in the above example that once the minute value is more than 60, we subtract 60 from that column and add 1 to the hour. In this case, we got 102 minutes in the minutes' column which is more than 60 and so we subtracted 60 from 102, making the final answer in the minutes' column as 102 − 60 = 42 minutes. Moreover, in order to adjust it, we added 1 in the hours' column.

The method I shall discuss here will make the whole process of adding/subtracting numbers related to time easy as you need not add subtract multiples of 60 in the minutes' column and add 1 in the hours' column.

Rules:

- Simply add the two numbers without considering it to be an hour–minute problem.
- Add 40 to the final result.

Example: Add 4 hours 34 minutes and 5 hours 57 minutes.
Solution: First write 434 and 557 for 4 hours 34 minutes and 5 hours 57 minutes respectively. Add them as shown below:

434 + 557 = 991

Add 40 to the sum of the two, i.e. 991 + 40 = 1031

Hence, 4 hours 34 minutes + 5 hours 57 minutes = 10 hours 31 minutes

Example: Add 6 hours 44 minutes and 12 hours 39 minutes.
Solution: First write 644 and 1239 for 6 hours 44 minutes and 12 hours 39 minutes. Add them—644 + 1239 = 1883

Add 40 to the sum, i.e. 1883 + 40 = 1923

Hence, 6 hours 44 minutes + 12 hours 39 minutes = 19 hours 23 minutes

Subtracting Time

Using subtraction with numbers related to time (Hours–Minutes) is as simple as adding them. Follow these three simple steps and you will get the answer.

a) First subtract the hours' bit separately.
b) Subtract the minutes' part in the same manner. Don't worry if you get a negative sign. It may be that you have to subtract 17 from 12.

c) If the result of the minutes' side of it comes out in the negative, then add 60 there and to balance the result, subtract 1 from the hours' part.

Example: Subtract 2 hours 27 minutes from 4 hours 34 minutes.
Solution: a) First subtract the hours' part: 4 − 2 = 2
b) Then subtract the minutes' part: 34 − 27 = 07 minutes

Hence, 4 hours 34 minutes − 2 hours 27 minutes = 2 hours 07 minutes

Example: Subtract 2 hours 47 minutes from 4 hours 14 minutes.
Solution: a) First subtract the hours' part: 4 − 2 = 2
b) Now subtract the minutes' part: 14 − 47 = −33 minutes

Since, the minutes' part result is negative, add 60 to it, i.e. −33 + 60 = 27 minutes

Subtract 1 from the hours' part, i.e. 2 − 1 = 1 hour

Hence, 4 hours 14 minutes − 2 hours 47 minutes = 1 hour 27 minutes

You can do subtraction with the method we used in addition. Simply subtract the two numbers removing the hours and minutes heads. If the minutes' part (last 2 digits) is more than 60, then subtract 40 to get the correct answer.

Example: Subtract 2 hours 47 minutes from 4 hours 14 minutes.
Solution: Write 414 and 247 in place of 4 hours 14 minutes and 2 hours 47 minutes respectively.

Subtract as usual, i.e. 414 − 247 = 167

Since the last two digits is 67 > 60, subtract 40 from the final result.

167 = 1 hour + (67 − 40) = 1 hour 27 minutes

Example: Subtract 13 hours 27 minutes from 17 hours 04 minutes.
Solution: Write 1704 and 1327 in place of 17 hours 04 minutes and 13 hours 27 minutes respectively

Subtract as usual = 1704 − 1327 = 377

Since the last two digits comprise 77, that is, it is more than 60, subtract 40 from the final result.

So, 377 = 3 hours + (77 − 40) = 3 hours 37 minutes

Practice Problems

Solve the following:

a) 4 hrs 56 mins + 11 hrs 29 mins
b) 12 hrs 29 mins + 8 hrs 59 mins
c) 6 hrs 38 mins + 5 hrs 44 mins
d) 11 hrs 27 mins + 14 hrs 47 mins
e) 12 hrs 48 mins − 7 hrs 54 mins
f) 14 hrs 12 mins − 9 hrs 48 mins
g) 2 hrs 37 mins − 1 hrs 41 mins
h) 5 hrs 44 mins − 3 hrs 51 mins

II
MULTIPLICATION, BUT WITHOUT TEARS

It Is No Puzzle

Multiplication is one mathematical operation that can appear daunting and can sometimes make you feel annoyed. Such a situation arrives when you sometimes get wrong answers repeatedly.

Let's begin the chapter with a puzzle.

Three friends were returning from college when they saw a speeding car hitting a cyclist while he was crossing the road. The car didn't stop after hitting the cyclist and drove away leaving the man yelling on the road.

They shouted, desperate for help, but no passerby responded. In the meantime, they noticed the registration number of the car and tried to remember it in their own way. They rushed the person to the nearby hospital.

The doctor called the police as this was a road accident and a case was registered. In order to do a proper investigation, the police started questioning the three friends and asked them if they could provide the registration number of the car so that the driver who had hit the person could be caught.

All three friends tried a lot to remember the car number but failed. Thus they decided to zero in upon the number of the car as they had remembered it with some mathematical properties behind it.

First student: Sir, the first two digits of the car number were the same.

Second student: Sir, the last two digits of the car number were same.

Third student: The four digits made for a square number.

Can you guess the car number to help the police nab the guilty?

Solution: The car registration number is a four-digit number with the first two digits same and also the last two digits same. If the first two digits are **a** and last two digits **b**, then

Number = 1000 a + 100 a + 10 b + b
= 1100 a + 11 b
= 11 (100a + b)

This clearly shows that the four-digit number is divisible by 11. The policeman was a lover of mathematics so he finally cracked the problem and knew the registration number of the car and later caught the driver.*

Since it is a perfect square number, it will also be divisible by $11^2 = 121$.

Remember:

- A number is not a perfect square if it ends with 2, 3, 7 or 8.
- A perfect square will end with 0, 1, 4, 5, 6 and 9.

Hence **b = 0, 1, 4, 5, 6, and 9**. This implies **a = 11, 10, 7, 6, 5 or 2**

Since a = 11 or 10 is ruled out, we will have a = 7, 6, 5 or 2.

a	b	Four-digit number
7	4	7744
6	5	6655
5	6	5566
2	9	2299

Out of the 4 options, we have 7744—a perfect square.

Hence, car registration number was $7744 = (88)^2$

Now let's begin with number 11.

Most of the times, while solving mensuration problems, you

---

*For the divisibility rule of 11, see Section III on division.

must have used the value of π as 22/7 and multiplied that with 22. Let's explore it with an easier method. You can write 22 = 2 × 11. Learning this technique will surely help in mensuration as well.

Rule:

- Place the number to be multiplied by 11 in a bracket and put zeros on either side.
- Start adding the two numbers one at a time from right to left. If the sum of two numbers in any case exceeds 10, the digit at the tenth place shall be carried over to the next sum, as is usually done in simple addition.

Example: Multiply 16 by 11.
Solution:
Place the number in a bracket and put zeros on either side.

Add the digits from the right to the left as shown above.
0 + 1 | 1 + 6 | 6 + 0
= 1 7 6
Hence 16 × 11 = 176

Example: Multiply 4876254 by 11.
Solution:
Place the number in a bracket and put zeros on either side.

Add the digits from the right to the left as shown above.
= 0 + 4 | 4 + 8 | 8 + 7 | 7 + 6 | 6 + 2 | 2 + 5 | 5 + 4 | 4 + 0
= 4 | 12 | 15 | 13 | 8 | 7 | 9 | 4
= 4 | 12 | 16 | 3 | 8 | 7 | 9 | 4
= 4 | 13 | 6 | 3 | 8 | 7 | 9 | 4
= 5 3 6 3 8 7 9 4
Hence, 4876254 × 11 = 53638794

Friends, 11 can also be written as 10 + 1. Let's take one example to understand the multiplication of a number by 11 with this technique.

Example: Multiply 15 by 11.
Solution: $15 \times 11 = 15 \times (10 + 1)$
$= 150 + 15 = 165$

This means that if you have to multiply a number by 11, simply put a zero at the end of the number and then add the original number to it.

Example: Multiply 1275 by 11.
Solution: Add 0 at the end of 1275 = 12750
Now add the number to it = 12750 + 1275 = 14025

Multiplication with Number 9

9 is the largest one-digit number and the loveliest number in mathematics. You will admit that too when you would have finished turning the last page of this book. But let's first explore the distributive property of a number.

If a, b and c are three numbers then,
$$a \times (b \pm c) = a \times b \pm a \times c$$
10 = 9 − 1

Example: Multiply 15 by 9.
Solution: $15 \times 9 = 15 \times (10 - 1)$
$= 15 \times 10 - 15 \times 1$
$= 150 - 15 = 135$

This can be further simplified in two lines:
a) Add zero at the end.
b) And finally subtract the number from the product.

Example: 146×9
Solution: $146 \times 9 = 146 \times (10 - 1)$
$= 1460 - 146 = 1314$

Example: Multiply 5784 by 9.
Solution: 5874 × 9 = 57840 − 5784 = 52056

I hope you enjoyed this method? Why not make this multiplication even easier?

a) Subtract 1 from the multiplicand.
b) Now subtract the truncated number after removing the last digit.
c) Finally, write the complement of the last digit removed from 10 at the end.

Example: Multiply 16 by 9.
Solution: a) Subtract 1 from the multiplicand, i.e. 16 − 1 = 15
b) Now subtract the truncated number after removing the last digit, i.e. 15 − 1 = 14 (5 is removed, so truncated digit is 1)
c) Finally, write the complement of the last digit (6) removed from 10 at the end = 4
Hence, 16 × 9 = 144

Example: Multiply 124 by 9.
Solution: a) Subtract 1 from the multiplicand, i.e. 124 − 1 = 123
b) Now subtract the truncated number after removing the last digit, i.e. 123 − 12 = 111
c) Finally, write the complement of the last digit removed from 10 = 10 − 4 = 6
Hence, 124 × 9 = 1116

Example: Multiply 456789 by 9.
Solution: a) Subtract 1 from the multiplicand, i.e. 456789 − 1 = 456788
b) Now subtract the truncated number after removing the last digit = 456788 − 45678 = 411110
c) Finally, write the complement of the last digit (9) removed from 10 at the end = 10 − 9 = 1
Hence, 456789 × 9 = 4111101

Multiplication with 15

Multiplication of any number by 15 can be done by the previous method but I shall make some changes in it and help you learn another method which is better than the previous one.

Rule: a) Put a zero at the end of the multiplicand.
        b) Divide the previous result by 2.
        c) Add both the results.

Example: Multiply 12 by 15.
Solution: a) Put one zero at the end of 12: 120
            b) Divide it by 2: $120 \div 2 = 60$
            c) Add the two previous results: $120 + 60 = 180$
               Hence, $12 \times 15 = 180$

Example: Multiply 154 by 15.
Solution: a) Put one zero at the end of 154: 1540
            b) Divide it by 2: $1540 \div 2 = 770$.
            c) Add the two previous results: $1540 + 770 = 2310$

Example: Multiply 4569 by 15.
Solution: a) Put one zero at the end of 4569: 45690
            b) Divide it by 2: $45690 \div 2 = 22845$
            c) Add the two previous results: $45690 + 22845 = 68535$

Multiplication of Two-digit Numbers Below 50

Multiplication of two numbers can be done in a few seconds. First, see the process below.

Rule: a) Multiply first number by the tens' digit of the second number.
        b) Add zero at the end of the previous result.
        c) Multiply first number by the unit digit of the second number.

d) Add the two previous steps.

Example: Multiply 24 by 28.
Solution: Multiply 24 by 2: 24 × 2 = 48
Add zero at the end: 480
Multiply 24 by 8: 24 × 8 = 192
Add the two previous results: 480 + 192 = 672

Example: Multiply 36 by 47.
Solution: Multiply 36 by 4: 36 × 4 = 144
Add zero at the end: 1440
Multiply 36 by 7: 36 × 7 = 252
Add the two previous results: 1440 + 252 = 1692

Example: Multiply 46 by 48.
Solution: Multiply 46 by 4: 46 × 4 = 184
Add zero at the end: 1840
Multiply 46 by 8: 46 × 8 = 368
Add the two previous results: 1840 + 368 = 2208

Multiplication of the two smaller numbers can be done using the Vedic Mathematics technique.[†]

Case 1: If both numbers are between 10 to 20

In such a case, use the Nikhilam method to solve the problem.

a) First take the base 10 and write the differences against the number.
b) Do crosswise operation.
c) Multiply the differences.
d) If there is more than one digit on the right side, shift the leftmost digit to the left side so that only one digit is left on the right side.

[†]See my book *The Essentials of Vedic Mathematics*.

Example: 12 × 13
Solution: 12 + 2
   13 + 3
   ―――――
   15/   (Diagonal Sum = 12 + 3 or 13 + 2)
Multiply the differences: 2 × 3 = 6
Place the product on the right side.
Hence, 12 × 13 = 156

Example: 16 × 19
Solution: 16 + 6
   19 + 9
   ―――――
   25/   (Diagonal Sum = 16 + 9 or 19 + 6)
Multiply the differences: 6 × 9 = 54
Place the product on the right side.
16 × 19 = 25/54 (Base has more than 1 digit)
        = 25 + 5/4 (Leftmost digit on the right side is transferred to the left part)
        = 304

Case 2: Multiplication when both numbers are near 50 (40 to 60)

Here too, the same process as stated above will work. The main difference here, however, is that you have to divide the left part by 2 as base 50 = 100/2

Example: 42 × 48
Solution: 42 − 8
   48 − 2
   ―――――
   40 / 16 (Diagonal Sum/Difference = 42−2 = 40 or 48−8 = 40)
Divide the left part by 2: 40/2 = 20
Multiply the differences: 2 × 8 = 16
Place it on the right side.
Hence, 42 × 48 = 2016

Example: 54 × 59
Solution: 54 + 4
  59 + 9
  ─────
  63/   (Diagonal Sum = 54 + 9 = 59 + 4 = 63)
Divide the left side by 2: 63/2 = 31½
Multiply the differences: 4 × 9 = 36
Place it on the right side.
Hence, 54 × 59 = 31½/36
              = 31/50 + 36
              = 3186

Example: 56 × 47
Solution: 56 + 6
  47 − 3
  ─────
  53/ −18   (Diagonal Sum = 56 − 3 = 53)
Multiply the differences: 6 × −3 = −18
Divide the left side by 2: 53/2 = 26½
Hence, 56 × 47 = 26½/−18
              = 26 / 50 − 18
              = 2632

Case 3: Multiplication of numbers near base 100, 1000, 10000, etc.

Example: 104 × 109
Solution: 104 + 4
           × 109 + 9
           ─────────
             113 / 36
Diagonal sum: 104 + 9 = 109 + 4 = 113
Multiplication of difference: 4 × 9 = 36
Hence, 104 × 109 = 11336

Example: 94 × 89
Solution: 94 − 6
           × 89 − 11
           ─────────
             83 / 66

Diagonal sum: 94 − 11 = 83
Multiplication of difference: 6 × 11 = 66
Hence, 94 × 89 = 8366

Example: 84 × 99
Solution:
$$\begin{array}{r} 84 - 16 \\ \times\,99 - 1 \\ \hline 83\,/\,16 \end{array}$$

Diagonal sum: 84 − 1 = 83
Multiplication of difference: 16 × 1 = 16
Hence, 84 × 99 = 8316

Example: 92 × 99
Solution:
$$\begin{array}{r} 92 - 8 \\ \times\,99 - 1 \\ \hline 91\,/\,08 \end{array}$$

Diagonal sum: 92 − 1 = 91

Multiplication of difference: 8 × 1 = 08

(Number of digits on the right side depends upon the number of zeros in base. Here, base = 100 has two zeros, so 8 is written as 08)

Example: 84 × 99
Solution:
$$\begin{array}{r} 84 - 16 \\ \times\,99 - 1 \\ \hline 83\,/\,16 \end{array}$$

84 is 16 less than base 100, and 99 is 1 less than 100.
Diagonal sum: 84 − 1 = 83
Multiplication of difference: 16 × 1 = 16
Hence, 84 × 99 = 8316
This can be simplified further—

If the number to be multiplied is near any base (10, 20, 30, 40, 50,...100, 1000...), simply follow the algebraic rule.

$$(y + a)(y + b) = (y + a + b) \times y + ab$$

Example: 34 × 37 = ?
Solution: 34 and 37 both have the same digits in the tens' places.
$$34 \times 37 = (30 + 4)(30 + 7)$$
$$= (34 + 7) \times 30 + (4 \times 7)$$
$$= 1230 + 28$$
$$= 1258$$

Example: 67 × 69 = ?
Solution: 67 and 69 have the same digits in the tens' places.
$$67 \times 69 = (60 + 7)(60 + 9)$$
$$= (67 + 9) \times 60 + (7 \times 9)$$
$$= 4560 + 63$$
$$= 4623$$

Example: 117 × 103 = ?
Solution: 117 × 103 = (100 + 17)(100 + 3)
$$= (117 + 3) \times 100 + (17 \times 3)$$
$$= 12000 + 51$$
$$= 12051$$

Product of Number When Units' Digits Are Equal

Let us first understand the algebra behind such a multiplication.
$$(x + a)(y + a) = xy + ax + ay + a \times a$$
$$= xy + a(x + y) + a \times a$$

Example: 45 × 35
Solution: Here 45 and 35 both end with 5, i.e. units' digits are the same.
Hence, $45 \times 35 = 4 \times 3 \, / \, (4 + 3) \times 5 \, / \, 5 \times 5$
$$= 12 \, / \, 35 \, / \, 25$$
$$= 12 + 3 \, / \, 5 + 2 \, / \, 5$$
$$= 1575$$

Shift the digit to the extreme left of each product to the previous column. The calculation has to be completed from right to left keeping only a single digit in each separate column except in the extreme left one.

Example: 57 × 37

Solution: Here 57 and 37 both end with 7, i.e. unit digits are same.

Hence, 57 × 37 = 5 × 3 / (5 + 3) × 7 / 7 × 7
     = 15 / 56 / 49
     = 15 + 5 / 6 + 4 / 9
     = 20 / 10 / 9
     = 2109 (1 is shifted to the left)

Example: 104 × 204

Solution: 104 and 204 both end with 4.

104 × 204 = 10 × 20 + (10 + 20) × 4 + 4 × 4
    = 200 / 120 / 16
    = 200 / 120 + 1 / 6
    = 200 + 12 / 1 / 6
    = 21216

Example: 64 × 44

Solution: 64 and 44 both end with 4.

64 × 44 = 6 × 4 + (6 + 4) × 4 + 4 × 4
   = 24 / 40 / 16
   = 28 / 0 + 1 / 6
   = 28 / 1 / 6
   = 2816

Multiplication of Two/Three Digit-Numbers When Tens'/ Tens'-Hundreds' Digits Are The Same

Let's use the concept of algebra to explore how this method works.

We know: $(x + a)(x + b) = x^2 + (a + b)x + ab$

Multiplication of two-/three-digit numbers having the same figure in the tens' and hundreds' places, such as 26 × 28 or 402 × 409.

- Multiply the unit digit first = a × b
- Multiply the remaining digit with the sum of digits = x (a + b)
- Multiply the tens' digit / remaining digits except unit = x × x
- Add the above steps

Example: Multiply 87 by 86.
Solution:

- Multiply the unit digit first: 7 × 6 = 42
- Multiply the remaining digit with the sum of digits: x (a + b) = 8 (7 + 6) = 104 tens = 1040
- Multiply the tens' digit / remaining digits except unit: x × x = 8 × 8 = 64 hundreds = 6400
- Add 87 × 86 = 6400 + 1040 + 42 = 7482

Example: Multiply 287 by 284.
Solution:

- Multiply the unit digit first = 7 × 4 = 28
- Multiply the remaining digit with the sum of digits = x (a + b) = 28 (7 + 4) = 308 tens = 3080
- Multiply the tens' digit/remaining digits except unit = x × x = 28 × 28 = 784 hundreds = 78400
- Add = 287 × 284 = 78400 + 30800 + 28 = 81508

Example: Multiply 259 by 257.
Solution:

- Multiply the unit digit first = 9 × 7 = 63
- Multiply the remaining digit with the sum of digits = x (a + b) = 25 (9 + 7) = 400 tens = 4000

- Multiply the tens' digit / remaining digits except unit = $x \times x = 25 \times 25 = 625$ hundreds = 62500
- Add = $259 \times 257 = 62500 + 4000 + 63 = 66563$

Let's take another very interesting case of multiplication. You wouldn't have seen such a beautiful pattern of multiplication before! Multiplying any two numbers that have a number of zeros is easy. Suppose you are asked to multiply 700×900, then you will directly write the answer as 630000 as it involves simple multiplication. Moreover, you can use the same technique in multiplying two different numbers by increasing or decreasing the number such as to produce a multiple of tens in each case.

If x and y are two numbers and both are increased with a and b, such that x and y are now $x + a$ and $y + b$, then,

$xy = x(y + b) - (x + a)b + ab$
$= xy + xb - xb - ab + ab$

Moreover, it can also be written as—

$xy = y(x + a) - a(y + b) + ab$
$= yx + ya - ay - ab + ab$

Let's see an instance.

Example: Multiply 687 by 893.

Solution: Here, $x = 687$ and $y = 893$

By adding 13 to 687, it can be changed into a suitable number that is a multiple of 10. Similarly, add 7 in 893 to make it 900.

$687 + 13 = 700$
$x + a = x + a$
$893 + 7 = 900$
$y + b = y + b$

Applying the formula:
$xy = x(y + b) - (x + a)b + ab$
$687 \times 893 = 687 \times 900 - 700 \times 7 + 13 \times 7$
$= 618300 - 4900 + 91$
$= 613491$

You can do this by using the formula:
$$xy = y(x + a) - a(y + b) + ab$$
$$687 \times 893 = 893 \times 700 - 13 \times 900 + 91$$
$$= 625100 - 11700 + 91$$
$$= 613491$$

Example: Multiply 442 by 778.
Solution: $400 = 442 - 42 = x - a$
$800 = 778 + 22 = y + b$
Hence, 442×778 can be done using the method above:
$$xy = x(y + b) - b(x - a) - ab$$
$$442 \times 778 = 442 \times 800 - 22 \times 400 - 42 \times 22$$
$$= 353600 - 8800 - 924$$
$$= 343876$$

Multiplication of Two-digit Numbers by Reversing the Second Number

If you get the result of multiplying two numbers using an unconventional method and by reversing the second number, that would be amazing, isn't it? Let's first learn the placement of the numbers.

Example: Multiply 15 by 16.
Solution: Here we shall reverse the multiplier 16 and multiply 15 by 61 by placing 61 below 15 in three different positions.

```
     1 | 5          1    5 |        1   5 |
  × 6   1↓        ×↓6    1↓        × 6↓  1
  ─────────      ─────────────    ────────────
     1              6 + 5            30
```

$$= 1 \: / \: 11 \: / \: 30$$
$$= 1 + 1 \: / \: 1 + 3 \: / \: 0$$
$$= 240$$

Example: Multiply 77 by 49.
Solution: Reverse 49 and write 94 below 77 in three different positions

$$\begin{array}{ccc} 7\!\mid\!7 & 7\!\mid\quad 7\!\mid & 7\quad 7\!\mid \\ \times 9\ \ 4\!\blacktriangledown & \times 9\!\blacktriangledown\ \ 4\!\blacktriangledown & \times 9\!\blacktriangledown\ \ 4 \\ \hline 28 & 63\ +\ 28 & 63 \\ =28\ \ /\ \ 91\ \ /\ \ 63 & & \end{array}$$

$$= 3773$$

Example: Multiply 63 by 98.
Solution: Reverse the multiplier 98 and place as shown in the above examples.

$$\begin{array}{ccc} 6\!\blacktriangle\!3 & 6\!\blacktriangle\ \ 3\!\blacktriangle & 6\quad 3\!\blacktriangle \\ \times 8\ \ 9\!\mid & \times 8\!\mid\ \ 9\!\mid & \times\ \ 8\!\mid\ \ 9 \\ \hline 54 & 48\ +\ 27 & 24 \\ =54\ \ /\ \ 75\ \ /\ \ 24 & & \end{array}$$

$$= 6174$$

Now the question is how does this method work.

Let's first solve this: $(a + b) \times (c + d) = ad + ac + bd + bc$
Now, place the multiplier as directed above.

$$\begin{array}{ccc} a\ \ b & a\ \ b & a\ \ b \\ \times\ c\ \ d & \times\ c\ \ d & \times\ \ c\ \ d \\ \hline ad & ac\ +\ bd & bc \end{array}$$

As you can see, we get the same result after reversing the order of the multiplier and placing them in the position as shown.

This method reminds me of the Vedic techniques of multiplication that make use of the dot and cross method. Let's see two examples. For 2 × 2 multiplying in the Vedic way you need to understand the pattern of dots and crosses first.

Multiplication of two-digit numbers using Vedic techniques

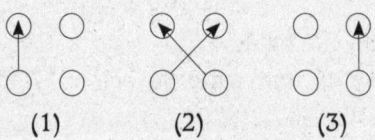

(1)　　　　(2)　　　　(3)

Example: Multiply 76 by 42.
Solution:

The arrangement of numbers is done on the top and bottom of the dots.

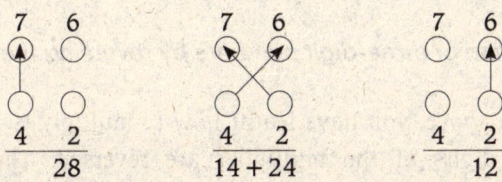

Arranging the numbers and adding them from right to left, taking only one digit at a time, we get the final result.

= 28 | 38 | 12

= 3192

This can be done by placing the numbers one below another but shifting the number in each column as shown here.

```
  2 8
    3 8
    + 1 2
  -------
  3 1 9 2
```

Example: Multiply 63 by 48.
Solution: First place the number on the dots and do the multiplication.

```
  6   3          6   3          6   3
  ↑   ○          ↘   ↙          ○   ↑
  ○   ○          ○   ○          ○   ○
  4   8          4   8          4   8
   24            48 + 12          24
```

= 24 | 60 | 24

= 3024

You can add the results in the following ways.

```
  2 8
  6 0
+ 2 4
-------
3 0 2 4
```

Multiplication of three-digit numbers by reversing the multiplier

In the case above, you have learnt how to multiply two numbers when the digits of the multiplier are reversed. The diagram above tells the reason why this method is valid. In the 2 by 2 multiplication method, the number of steps come out to be 3. In case of 3 by 3 multiplication scenarios, the number of operations to be carried out is 5.

Let's first multiply two numbers, say ABC and DEF, in their extended forms.

$(A + B + C) \times (D + E + F) = AD + AE + AF + BD + BE + BF + CD + CE + CF$

Now, change the order of the second from DEF to FED and see the arrangement done below. You will find that the same result is obtained in the end.

```
     A B C |  A B C | A B C | A B C | A B C
     |     |  |  |  | | | | |   | | |     |
F E D     |F E D   |F E D  |F E D   |     F E D
  AD         AE+BD   AF+BE+CD  BF+CE        CF
```

Example: Multiply 124 by 832.

Solution: First reverse 832 to get 238 and then place 2, 3 and 8 in the following way to get the exact answer.

```
   1 2 4      1 2 4       1  2  4     1 2 4        1 2 4
       |      |   |       |  |  |     | |              |
 2 3 8      2 3 8        2  3  8     2 3 8          2 3 8
 -----      ---------    ---------   ---------      -----
   8         3+16=19     2+6+32=40    4+12=16         8
```

$= 8/19/40/16/8$
$= 8/19/40+1/6/8$
$= 8/19+4/1/6/8$
$= 8+2/3/1/6/8$
$= 103168$

Example: Multiply 569 by 632.
Solution: First reverse 632 and multiply it.

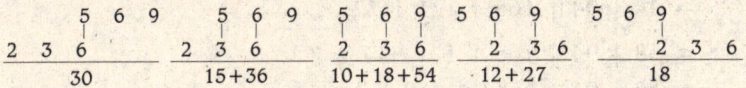

= 30/51/82/39/18
= 30/51/82/39+1/8
= 30/51/82+4/0/8
= 30/51+8/6/0/8
= 30+5/9/6/0/8
= 35/9/6/0/8
= 359608

Let's check out a 3 by 3 multiplication scenario using the dot and stick method.

Example: Multiply 548 by 159.
Solution:

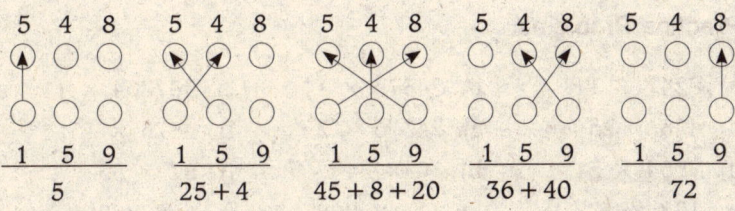

Separate each product with a small vertical rule, as shown below.
= 5 | 29 | 73 | 76 | 72
= 87132

Example 51: Multiply 659 by 898.
Solution: Arranging the numbers on the dots.

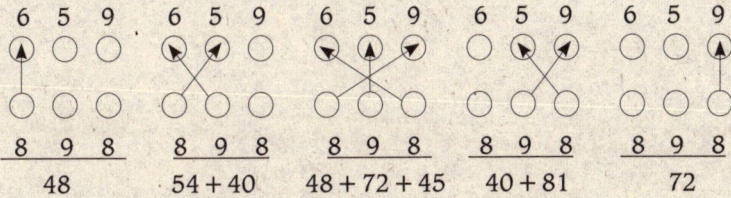

Arrange each of the products such that you separate them with a vertical rule as shown below.

= 48 | 94 | 165 | 121 | 72
= 48 | 94 | 165 | 121 + 7 | 2
= 48 | 94 | 165 | **12** 8 | 2
= 48 | 94 | 165 + 12 | 8 | 2
= 48 | 94 | **17** 7 | 8 | 2
= 48 | 94 + 17 | 7 | 8 | 2
= 48 | **11** 1 | 7 | 8 | 2
= 48 + 11 | 1 | 7 | 8 | 2
= **59 | 1 | 7 | 8 | 2**
= **591782**

Hence, 659 × 898 = 591782

Keep experimenting and enjoy multiplication!

Practice Problems

a) 8247 × 11
b) 24876 × 11
c) 987204 × 11
d) 428 × 25
e) 24376 × 25
f) 4876 × 15
g) 7654 × 51
h) 36 × 34
i) 87 × 83
j) 128 × 122
k) 112 × 998
l) 688 × 988
m) 107 × 95
n) 9997 × 9998
o) 252 × 248
p) 148 × 149
q) 506 × 494
r) 2487 × 9999

Multiply the following by reversing the digits in the multiplier:

a) 45 × 98
b) 86 × 59
c) 29 × 76
d) 112 × 872
e) 546 × 756

Multiplication Techniques from Around the World

Multiplication is being taught since primary school but the fear of making a mistake looms large over students' minds. There are around ten methods in Vedic Mathematics and those can help you multiply two numbers in just a matter of seconds and those I had mentioned in my book on Vedic mathematics. Moreover, there is the Trachtenberg System of multiplication which is widely used in European countries for speedy calculations. But apart from that, there are many other methods that can come to your aid to multiply two numbers in no time. In this chapter, you will find the multiplication methods from Russia, Japan, China, Egypt and many other countries. So here's a taste of the different methods of multiplication practised in different parts of the world.

The Russian Method of Multiplication

The Russian Peasant Method of multiplication is one of the tools used to multiply two numbers. It is said to be used by peasants in some parts of Russia. I won't say this is an effective multiplication method but it is fun. In ancient times, farmers used to calculate with the help of pebbles and so it is obvious that they must be using a single digit or two-digit numbers to multiply. However, this process can also be very effective with larger amount of numbers too, but in that case, the application becomes tough.

Let's enjoy doing mathematics!*

Rules:

- Make two columns.
- Write each of the two numbers at the top of each column.
- Double the number in the first column, and halve the number in the second column.
- If the number in the second column is odd, divide it by two and drop the remainder.
- If the number in the second column is even, cross out that entire row.
- Repeat the process until you get 1 in the second column.
- Add up the remaining numbers in the first column. The total is the product of your original numbers.

Example: Multiply 45 by 26.
Solution:

Column 1	Column 2
~~45~~	~~26~~
90	13
~~180~~	~~6~~
360	3
720	1
1170	Total

Using the same process, here are a few more examples.

---

*Of the methods to be discussed in this chapter, I have not explained the process of multiplying using one's fingers. But that finds mention in my book *Maths Made Easy*. Your fingers can help you multiply digits—whether single or double, with ease. This method is, however, not practical to introduce here and hence my focus will be on some other multiplication tricks.

Example: Multiply 126 by 47.
Solution:

Column 1	Column 2
126	47
252	23
504	11
1008	5
~~2016~~	~~2~~
4032	1
5922	**Total**

Example: Multiply 56 by 94.
Solution:

Column 1	Column 2
~~56~~	~~94~~
112	47
224	23
448	11
896	5
~~1792~~	~~2~~
3584	1
5264	**Total**

You may ask whether 45 × 26 is the same as 26 × 45 when the Peasant Method is used. The answer is obviously 'yes' as commutative laws still exist in this case. Have a look at the first example.

Example: Multiply 45 by 26.
Solution:

Column 1	Column 2
26	45
~~52~~	~~22~~
104	11
208	5
~~416~~	~~2~~
832	1
1170	**Total**

As you can see here, we multiply the first column by 2 and divide the second column by 2. It means we are simply grouping the numbers with no difference.

$8 \times 12 = 4 \times 24 = 2 \times 48 = 96 \times 1$. As long as we get the odd number in the second column, the multiplication runs smoothly but the question is why we cut the rows having even numbers in the second column. This is one of the most important things to know. We shall learn the principle behind it later but let me take a few more examples to illustrate the working process of this method. This time I am not putting the numbers in the boxes as shown above. See the working first and thereafter I shall unearth the reason as to why this method of multiplication works.

Example: Multiply 14 by 27.
Solution: First make two columns—Multiplicand and Multiplier—and follow the rule described above.

Multiplicand (×2)	Multiplier (/2)	
14	27	
28	13	
~~56~~	~~6~~	(Cancel them as 56 and 6 are both even)
112	3	
224	1	
378		

Hence, $14 \times 27 = 378$

Example: Multiply 654 by 76.
Solution:

Multiplicand (×2)	Multiplier (/2)
654	76
1308	38
2616	19

Contd. on p. 51

Contd. from p. 50

5232	9
10464	4
20928	2
<u>41856</u>	1
49704	

This method is not tricky and is not fruitful when applied to quick calculations but the question then is how were the ancient people, despite having no idea of the present methods of multiplication, able to nevertheless solve multiplication problems. I do agree that at the time they would not have been in a hurry to find the answers nor would they have multiplied two bigger numbers as we are doing here, but once you learn that they were capable of using the **binary system of multiplication**, you will be much happier about their techniques of multiplication. Besides, this is also the basis of computer multiplication.

Example: Multiply 112 by 235.
Solution:

Multiplicand (×2)	Multiplier (/2)
112	235
224	117
448	58
896	29
1792	14
3584	7
7168	3
<u>14336</u>	1
26320	

As promised, I now reveal the basic fundamentals of this method. Some people call it the Russian Peasant Method, some other call it the Ethiopian way of multiplication and yet some others call it the Egyptian method of multiplication. Although the name hardly matters, its working principle will certainly amaze you.

Let's multiply 6 by 9 using the Peasant Method:

Column 1	Column 2
6	9
12	4
24	2
48	1

Arrange dots of 6 by 9

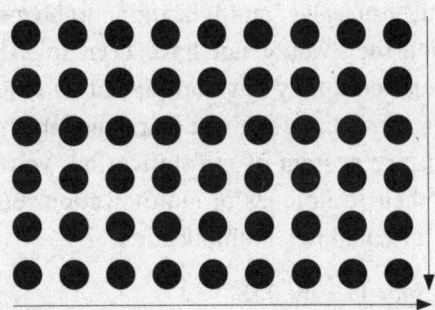

In the second row, we have multiplied the number in the first column by 2 but while dividing the number of the second column by 2, we haven't taken the remainder. Let's see the difference that occurred in the product of the 2 columns.

Column 1 = 6 × 9 = 54
Column 2 = 12 × 4 = 48
Difference = 54 − 48 = 6

In order to make the sum total, we rearrange the group as shown below.

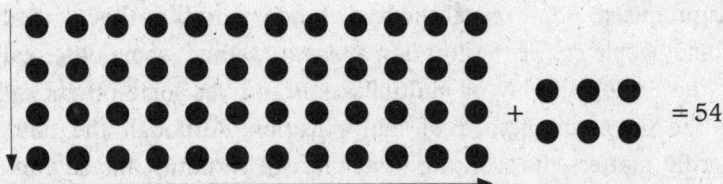

52 *Speed Mathematics*

This method basically does the same. It has a deep connection with the binary system. Let's check the aspect of the binary number. As you are aware, in the binary system, we divide a number by 2 and write the remainder down in a separate column. The remainder, when clubbed from the bottom to the top, gives the binary number of the corresponding decimal number.

Now, let's multiply 6 by 9 using the Peasant Method again but before that, convert 9 into a binary number.

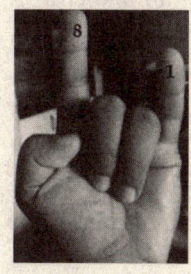

		R
2	9	
2	4	1
2	2	0
	1	0

8 + 1 = 9 in binary will represent 1001

Column 1	Multiply by 2	Column 2	Dividing by 2 (Remainder)
6	2^0	9	1
12	2^1	~~4~~	0
24	2^2	~~2~~	0
48	2^3	1	1

As you can see in the table above, whenever we get the remainder 0, we cancel the number in the corresponding row. I think the reason is clear now. Let's take one more example.

Example: Multiply 247 by 129.
Solution:

Column 1	Column 2
129	247
258	123
516	61

Contd. on p. 54

Contd. from p. 53

~~1032~~	~~30~~
2064	15
4128	7
8256	3
16512	1
31863	**Total**

This method will help make your multiplication of a number by 2 as well as division of a number by 2 simultaneously more effective. Though this method is not effective when doing fast calculations, but given that it has some connection to the binary number system in addition to it being used thousands of years ago by Russian and Egyptian farmers for calculations, it is worthwhile to note how advanced they were in computing simple multiplications using such binary techniques.

Practice Problems

Multiply the following:

a) 632 × 8
b) 3987 × 7
c) 254 × 16
d) 1256 × 65
e) 574 × 456
f) 8732 × 653
g) 777 × 623
h) 5697 × 6253
i) 632 × 937
j) 127 × 212

Grid Multiplication

The Grid Method is the best when it comes to teaching multiplication to children at the primary level. In this method, we decompose a larger number into small parts on the basis of their place value.

$34 = 3 \times 10 + 4 = 30 + 4$
$235 = 2 \times 100 + 3 \times 10 + 5$

This method will certainly help you understand the basic

composition of the number writing pattern. Let's see how it works.

- First make a box of 2 by 2 or 3 by 3
- Write the number in expanded form in each box.

Example: Multiply 36 by 29.
Solution:

×	30	6
20		
9		

Multiply the first row (top) by 20.

×	30	6
20	600	120
9		

Multiply the top digit by 9.

×	30	6
20		
9	270	54

Now add the four numbers: 600 + 120 + 270 + 54 = 1044

Example: Multiply 58 by 47.
Solution:

×	50	8
40		
7		

Multiply the first row (top) by 40.

×	50	8
40	2000	320
7		

Multiply the top digit by 7.

×	50	8
40		
7	350	56

Now add the four numbers: 2000 + 320 + 350 + 56 = 2726

Example: Multiply 457 by 342.
Solution:

$457 = 400 + 50 + 7$
$342 = 300 + 40 + 2$

×	400	50	7
300			
40			
2			

Multiply the top digit by 300, 40 and 2.

×	400	50	7	Add Row
300	120000	15000	2100	137100
40	16000	2000	280	18280
2	800	100	14	914

137100 + 18280 + 914 = 156294

Example: Multiply 8957 by 342.
Solution:

$8957 = 8000 + 900 + 50 + 7$
$342 = 300 + 40 + 2$

×	8000	900	50	7
300				
40				
2				

Multiply the top digits by 300, 40 and 2. Add each row separately and finally add the last column that has the total sum of each row.

×	8000	900	50	7	Add Row
300	2400000	270000	15000	2100	2687100
40	320000	36000	2000	280	358280
2	16000	1800	100	14	17914

2687100 + 358280 + 17914 = 3063294

Hope you have understood the process? Now, let's move to another method of multiplication. This is better known as the Lattice Multiplication Method or the Chinese Lattice Multiplication Method.

Chinese Lattice Multiplication Method

Lattice multiplication techniques have been in use since the thirteenth century but one is not sure who used this method for the first time. This technique of multiplication is quite interesting; however, it isn't easy to work with since it is time-consuming. Nevertheless, check the example and enjoy it!

Method:

- First draw a grid and split each cell diagonally.
- Write the multiplicand and multiplier on the top and right side/left side of the grid.
- Start with a digit of the multiplier and fill each cell of the lattice with the product of digits in the top row.
- Add the digits diagonally.
- The final answer appears in L shape.

Example: Multiply 24 by 82.

Solution: First make a 2 by 2 grid as shown below. Write multiplicand 24 at the top and multiplier 82 on the right side. Join each cell diagonally, from the left bottom and upwards to the right.

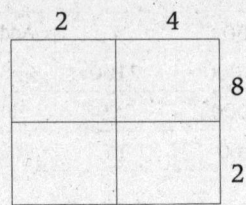
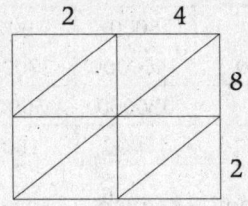

First multiply by the unit digit (2). Write the unit digit in the cell at the bottom and tens' digit at the top. Once the operation is done, multiply 24 by the tens' digit.

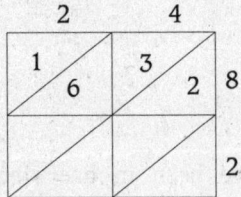
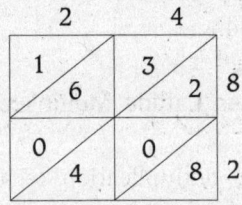

After all the cells are filled in this manner, the digits falling in each part of the diagonally dissected squares in terms of those that fall on a single line are summed; you work from the bottom right diagonal to the top left. Each sum is written where the diagonal ends. Make an L shape and the digits that fall on the L, comprise the final result.

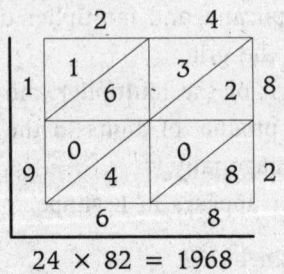

24 × 82 = 1968

Example: Multiply 56 by 314.
Solution: Make a table of 2×3 as there are 2 digits in the multiplicand and 3 digits in the multiplier. Join the ends of each square box diagonally as shown below. First multiply 56 by 3,

58 *Speed Mathematics*

then by 1 and 4, as done in the normal multiplication process. The unit digit of the product in each box will be filled in the bottom part of the diagonal. Once the boxes are filled, add the numbers falling on each of the diagonal arcs separately. Make the L shape and the numbers falling on the L comprise the final result.

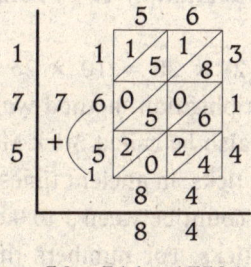

$56 \times 314 = 17584$

Example: Multiply 248 by 729.

Solution: Since it is a 3 by 3 multiplication scenario, make a 3 by 3 box. Place the multiplicand digit 248 at the top of the box and the multiplier 729 in the adjacent side of the box as shown. Fill the box as directed in the example above. Keep adding the diagonal elements and you will get the final answer lying along the L.

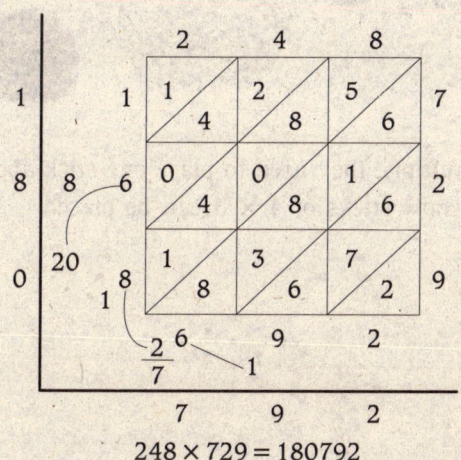

$248 \times 729 = 180792$

Multiplication Techniques from Around the World

Japanese Multiplication Technique

The Japanese multiplication technique is based on the principle of distributive property. For three numbers, a, b and c, we have,
$$a \times (b + c) = a \times b + a \times c$$

Suppose you have to multiply 12 by 25, then it can be written as
$$12 \times 25 = (10 + 2) \times 25 = 10 \times 25 + 2 \times 25 = 300$$

The Japanese multiplication method works in the same way.

This method may also be called Stick Method or Rod Method as it initially required sticks. In ancient times, when people started counting, they used to count between 1 to 9 either on their fingers or with the help of sticks. For numbers that were multiples of ten, they used small stones and for larger numbers, they used bigger stones.

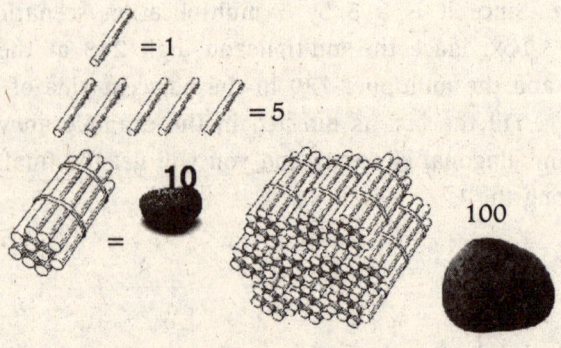

In order to multiply, they used to place one stick above another.

Let's see how sticks of 4 × 3 can be placed.

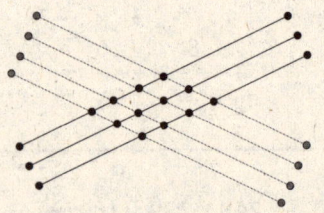

The point of intersection of sticks is then counted to get to the final answer. Here, 4 horizontal and 3 slanting lines are placed one above another to get 12 intersection points.

Hence, $4 \times 3 = 12$

Now let's see how it works.

1. Draw a set of parallel sticks representing the multiplicand. Each digit of the multiplicand will be represented with a gap.
2. Draw a set of parallel sticks representing multipliers. Each digit of the multiplier will be demonstrated with the help of sticks with a proper gap in between each one of them.
3. Put dots to mark intersection points.
4. On the left corner, put curved lines. Do the same on the right corner.
5. Count the points of intersection on the left, middle and the right corner.
6. If the number on the right is greater than 9 (i.e. it is a two-digit number) then the digits at the tens' place will be moved to the middle place and added with the number obtained in the middle. The same operation will be continued thereafter.
7. The final answer will be obtained by writing the points of intersection from left to right.

Example: Multiply 15 by 13.

Solution: 15 has two digits—1 and 5; so,

1. Draw 1 and 5 sticks vertically as shown below. The vertical sticks represent here the multiplicand. Similarly, 1 and 3 lines—the latter with gaps—are drawn horizontally to represent the multiplier.
2. Circle each corner (point of intersection at the left, right).
3. Count the number of intersection points in the middle.
4. Write the value of each intersection point. If the number

of intersection points is more than 9 then the digit at the tens' place will be added to the previous point of intersection value. Here the number of intersection points at the extreme right is 15 which is greater than 9; so, 1 at the tens' place will be moved to the point of intersection value in the middle. Hence, the point of intersection value of sticks in the middle will get increased by 1.

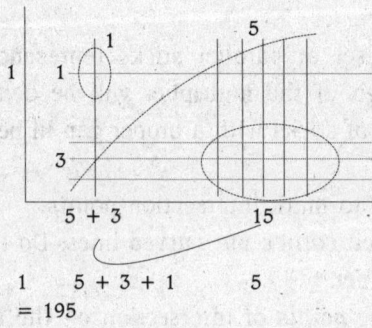

$$1 \quad 5+3+1 \quad 5$$
$$= 195$$

Example: Multiply 34 by 23.
Solution:

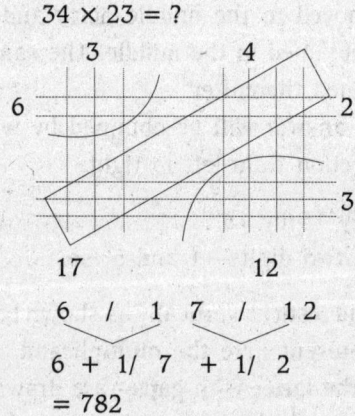

$$6 + 1/\ 7 + 1/\ 2$$
$$= 782$$

As you can see, the Japanese method is quite effective and easy to understand as far as the multiplication of two numbers is concerned. Now, let's multiply 2 three-digit numbers using the same method to check how effective it is.

Example: Multiply 234 by 333.
Solution:

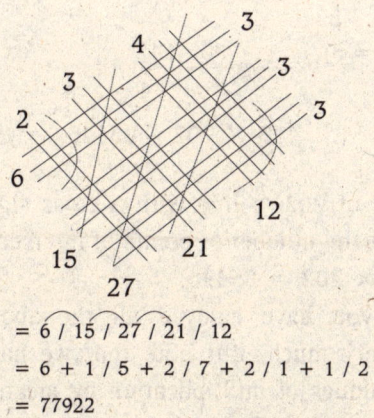

$= 6 / 15 / 27 / 21 / 12$
$= 6 + 1 / 5 + 2 / 7 + 2 / 1 + 1 / 2$
$= 77922$

Clearly, despite being easy, this method of multiplication can't be said to be very effective when faster calculations are required to be done. But before I move to the new method of multiplication, let's enjoy another instance of a 3 by 3 multiplication scenario with the Japanese method.

Example: Multiply 423 by 201.
Solution: In all the above examples you have seen that the multiplier has all digits except '0'. Let's consider a case when either the multiplicand or multiplier has at least one zero. In such a case, draw the line for each digit in the multiplicand and multiplier. You have to draw a line for that digit whose place value is zero. Cut the lines with cross marks at the top and the bottom. Now follow the rule directed in the above examples.

Multiplication Techniques from Around the World

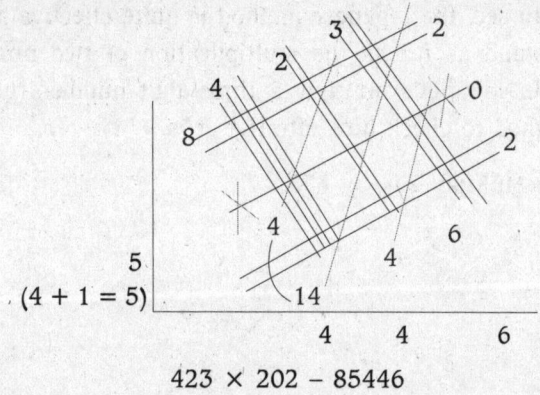

423 × 202 − 85446

Here, for the '0' of 202, a line with a cross sign at the top is used. Now, count the number of points of intersection diagonally.

Hence, 423 × 202 = 85446.

I do hope you have enjoyed all the above methods of multiplication very much. But note that we have learnt some innovative techniques of multiplication by means of which you will be able to calculate faster than ever before in the chapter before this! Keep practising!

Practice Problems

Apply on the following the Russian Peasant Method/Column Method and the Japanese method:

a) 14 × 15 b) 17 × 19 c) 96 × 94
d) 106 × 108 e) 94 × 108 f) 98 × 99
g) 995 × 987 h) 982 × 1007 i) 1004 × 1042

Multiply the following using only the Russian Peasant Method:

a) 632 × 8 b) 3987 × 7 c) 254 × 16
d) 1256 × 65 e) 574 × 456 f) 8732 × 653
g) 777 × 623 h) 5697 × 6253 i) 632 × 937
j) 127 × 212

Multiplication Using Napier's Rods

Multiplication can be done quite easily using Napier's Rods. One can find such rods in mathematics laboratories. Multiplication using rods is similar to students using the abacus. At the beginning, you will find it difficult to handle the rods but with some practice, you can get the result you desire. Before discussing how Napier's Rods work, let us get a little acquainted with Napier's life.

John Napier, a Scottish mathematician, was born in 1550 in Edinburgh, Scotland. His father, Archibald Napier was an important man of the sixteenth century. He is best known for his invention of logarithms. He also invented what came to be known as Napier's Rods. Napier's Rods or bones are ten oblong pieces of wood or other material with square ends. Each of the four faces of each rod contains multiples of one of the nine digits. The first rod contains multiples of 0, 1, 9 and 8, the second of 1, 2, 9, 7, the third of 0, 3, 6, 9, the fourth of 0, 4, 9, 5, the fifth of 1, 2, 8, 7, the sixth of 1, 3, 8, 6, the seventh of 1, 4, 8, 5, the eighth of 2, 3, 7, 6, the ninth of 2, 4, 7, 5 and the tenth of 3, 4, 6, 5. Each rod, therefore, contains on two of its faces multiples of digits that are complementary to those on the other two faces; and the multiples of a digit and its complement are reversed in position.

1	1	2	3	4	5	6	7	8	9
2	2	4	6	8	1/0	1/2	1/4	1/6	1/8
3	3	6	9	1/2	1/5	1/8	2/1	2/4	2/7
4	4	8	1/2	1/6	2/0	2/4	2/8	3/2	3/6
5	5	1/0	1/5	2/0	2/5	3/0	3/5	4/0	4/5
6	6	1/2	1/8	2/4	3/0	3/6	4/2	4/8	5/4
7	7	1/4	2/1	2/8	3/5	4/2	4/9	5/6	6/3
8	8	1/6	2/4	3/2	4/0	4/8	5/6	6/4	7/2
9	9	1/8	2/7	3/6	4/5	5/4	6/3	7/2	8/1

Napier's Rods are good for multiplying a long number by a single number but it can also be used to multiply a long number by another long number.

Example: Multiply 123456 by 7.

1	1	2	3	4	5	6
2	2	4	6	8	1/0	1/2
3	3	6	9	1/2	1/5	1/8
4	4	8	1/2	1/6	2/0	2/4
5	5	1/0	1/5	2/0	2/5	3/0
6	6	1/2	1/8	2/4	3/0	3/6
7	7	1/4	2/1	2/8	3/5	4/2
8	8	1/6	2/4	3/2	4/0	4/8
9	9	1/8	2/7	3/6	4/5	5/4

Arrange the rods as shown above and look at the numbers in the seventh row, starting from the right, and write down the numbers obtained by adding the digits in the parallelogram as shown.

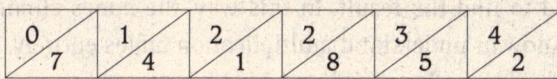

Now keep on adding the digits as discussed earlier in Vedic multiplication.

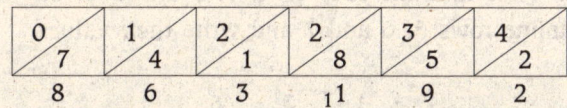

If the sum of digits in the respective column is more than 9, write the number exceeding in subscript as shown. Carry the digit written as a subscript to add into the next column. The final sum will look like as shown here.

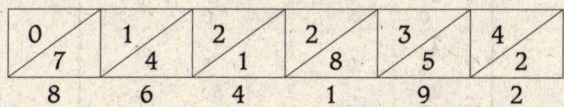

Therefore, the final result of multiplication is 2318128.

Example: Multiply 46258 by 7.

Solution: The seventh square of rod 8 contains the number 56; so, 56 is placed in the units position. The seventh square of rod 5 contains the number 35; so, 35 is placed in the tens' position. Continuing this process for the remaining rods, we get

$$
\begin{aligned}
4 \times 7 &= 28 \\
6 \times 7 &= 42 \\
2 \times 7 &= 14 \\
5 \times 7 &= 35 \\
\underline{8 \times 7} &= 56 \\
46258 \times 7 &= 323806
\end{aligned}
$$

Multiple-digit multipliers are processed the same way, digit by digit. These digits may be written down and shifted a column just as in ordinary pencil-and-paper arithmetic.

Once all the numbers have been multiplied, the partial sums are added to find the result. In this way, the bones eliminate the need to know or understand multiplication tables entirely, because multiplication has been replaced by a simple additive procedure.

Example: Multiply 4138 by 567.

Solution: First take the rods of 4, 1, 3 and 8 as shown here. Now examine rows 5, 6 and 7 and write their values.

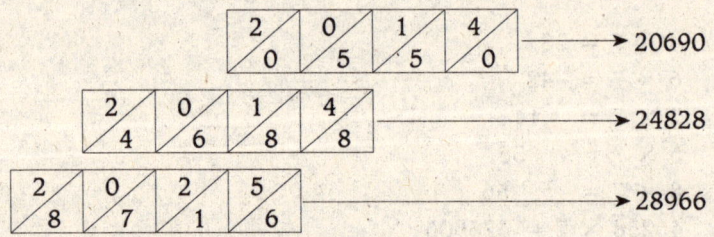

Write the multiplication values of 5, 6 and 7 as shown below. Add each diagonal separately and write the value adjacent to the box as shown here.

```
  2   0   1   4
    0   5   5   0    ──► 20690

  2   0   1   4
    4   6   8   8    ──► 24828

  2   0   2   5
    8   7   1   6    ──► 28966
```

Write the value of each box indicated by arrows according to the place value system as shown below and you have the final result at the end.

$$20690$$
$$24828$$
$$28966$$
$$\overline{2346246}$$

Example: Multiply 456 by 3579.
Solution:

456 × 3579 = ?

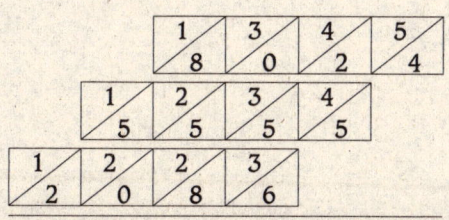

The result of 6 times 3579 = 21474
The result of 5 times 3579 shifted one place left = 178950
The result of 4 times 3579 shifted two places left = 1431600

```
Partial Sum together =    2 1 4 7 4
                        1 7 8 9 5 0
                      +1 4 3 1 6 0 0
                       ─────────────
                       1 6 3 2 0 2 4
```

This method works beautifully if you have the rods with you. In many schools, mathematics laboratories have been set up and beautiful demonstrations explaining the art of multiplying with the help of Napier's bones or rods can be conducted for children.

Practice Problems

Multiply the following using Napier's Rods:

a) 45632 × 8
b) 543987 × 7
c) 254 × 6
d) 3256 × 65
e) 5674 × 456
f) 8732 × 653
g) 777 × 623
h) 5697 × 6253
i) 632 × 937
j) 2574 × 9327
k) 5632 × 3589
l) 689 × 394

Multiplication Using Napier's Rods

III
DIVISION, DID YOU SAY 'INSOLUBLE'?

The Egyptian Method of Division

Insoluble? Not quite. Let me introduce you to ways that will help make division a far less fear-inducing arithmetical operation for you to deal with.

The Egyptian method of division, like the Egyptian method of multiplication, is based on the binary number system.

Here are the rules:

- Make two columns.
- Place the divisor in the second column.
- In the first column start with 1.
- Multiply the first and second column by 2.
- Ensure that the maximum value in the divisor column is less than the dividend.
- Add number in the second column so as to reach the dividend.
- Find the number in the first column corresponding to the number taken in the second column.

Example: Divide 13188 by 314.
Solution: Make two columns.

1	314	✗
2	628	✓
4	1256	✗
8	2512	✓
16	5024	✗
32	10048	✓

Since the dividend is 13188 which is less than 20096—the double of 10048—we will stop here. Now, add the number in the second column so as to get 13188.

10048 + 2512 + 628 = 13188

Hence, 13188 ÷ 314 = 32 + 8 + 2 = 42

Example: Divide 215919 by 297.
Solution: Make two columns.

1	297	✓
2	594	✓
4	1188	✓
8	2376	✗
16	4752	✓
32	9504	✗
64	19008	✓
128	38016	✓
256	76032	✗
512	152064	✓

297 + 594 + 1188 + 4752 + 19008 + 38016 + 152064 = 215919

Add the ticked column so as to get the exact dividend. The quotient will be the sum of the corresponding numbers in the first column.

1 + 2 + 4 + 16 + 64 + 128 + 512 = 727

Hence, 215919 ÷ 297 = 727

As you can see, here we are doubling the first part and the number we obtained in the first column is in the power of two.

Being time-consuming, this method therefore, is not very good. Here's another example to put my point more clearly.

Example: Divide 180 by 28.

Solution: Now make three columns instead of two.

	Dividend	Remainder	Quotient
1	28		☒
2	56	68 − 56 = 12	☑
4	112	180 − 112 = 68	☑
			2 + 4 = 6

What did I do here?

I multiplied the divisor—28 × 1 = 28; 28 × 2 = 56; 28 × 4 = 112, and put it in the first column. Since our next multiplication—28 × 8 = 224 > 180 (Dividend)—this is not taken into consideration.

Now add 28 + 56 + 112 > 180 and 56 + 112 = 168 < 180 so that only these two entries will give you the quotient and remainder. Moreover, 56 + 112 also doesn't make 180; so add the remainder column. Then, in the remainder column, start subtracting from the bottom list, i.e. 112.

180 − 112 = 68

68 − 56 = 12

12 − 28 = −16 is not possible, so remainder in this case is 12.

Quotient = 2 + 4 = 6 (Only ticked entries)

Example: Divide 301 by 17.

Solution: Now, make three columns.

	Dividend	Remainder	Quotient
1	17	29 − 17 = 12	√
2	34	×	
4	68	×	
8	136	×	
16	272	301 − 272 = 29	√
			1 + 16 = 17

The above examples are enough to show how sophisticated a technique such as the binary number system was being used by Egyptians at a time when the technique had not been explored fully. In fact, it is a technique based on which computers work.

Division by Repeated Subtraction

As you know, division is repeated subtraction. In fact, at the primary levels, we are taught this method. Let's see two examples.

a) Divide 16 by 4

```
 16
 −4      1st
 ──
 12
 −4      2nd
 ──
  8
 −4      3rd
 ──
  4
 −4      4th
 ──
  0
```

b) Divide 27 by 3

```
 27
 −9      1st
 ──
 18
 −9      2nd
 ──
  9
 −9      3rd
 ──
  0
```

Let's explore it with a bigger division.

Example: Divide 126 by 7.
Solution:

```
 7 | 146   | 10
    -70    |
    ────   |
     76    | 10
    -70    |
    ────   |
      6    | 20
```
Remainder

Quotient = 20; Remainder = 6

Example: Divide 1458 by 87.
Solution:

```
 87 | 1458  | 10
     -870   |
     ────   |
      588   |
     -435   | 5
     ────   |
      153   |
      -87   | 1
     ────   |
       66   | 16
```

Quotient = 16 and Remainder = 66

Example: Divide 1220 by 16.
Solution:

```
 16 | 1220  |
     -800   | 50
     ────   |
      420   |
     -320   | 20
     ────   |
      100   |
      -80   | 5
     ────   |
       20   |
      -16   | 1
     ────   |
        4   | 76
```

The Egyptian Method of Division

There is one more beautiful technique of division that is known as the Italian Method of Long Division but before I begin to explain you that, let me tell you about the divisibility rule of some of the numbers.

Divisibility Rule

2: A number is divisible by 2 if it ends with 0, 2, 4, 6 and 8
Example: 128, 3486, 245674, 452, 910—are divisible by 2 as they end with 8, 6, 4, 2, 0 respectively.

3: A number is divisible by 3 if the sum of the digits of the number is divisible by 3.
Example: 234, 4152, 67521—are divisible by 3.
    234: 2 + 3 + 4 = 9 is divisible by 3.
    4152: 4 + 1 + 5 + 2 = 12 is divisible by 3.
    67521: 6 + 7 + 5 + 2 + 1 = 21 is divisible by 3.

4: A number is divisible by 4 if it ends with 00 or the last two digits (as a single number) is divisible by 4.
Example: 124, 2468 and 456900 are divisible by 4.
    124: 24 is divisible by 4.
    2468: 68 is divisible by 4.
    456900: It ends with 00.

5: A number is divisible by 5 if it ends with 0 or 5.
Example: 2460 and 45895 are divisible by 5.

6: A number is divisible by 6 if it is divisible by 2 as well as 3.
Example: 42 is divisible by 6 as it is divisible by 2 (42 ends with 2) and 3 (4 + 2 = 6).

7: Double the last digit and subtract it from the remaining leading truncated number. If the result is divisible by 7, then the original number is divisible by 7. Apply this rule over and over again as necessary.

Example: Is 343 divisible by 7?
Solution: $34 - 2 \times 3 = 28$ is divisible by 7; so, 343 is divisible by 7.

Example: Is 1602 divisible by 7?
Solution: 1602: $160 - 2 \times 2 = 156$
And 156: $15 - 2 \times 6 = 3$
Therefore, 1602 is not divisible by 7.

8: A number is divisible by 8 if it ends with 000 or its last three digits (as a single number) is divisible by 8.
Example: 246000, 5681009128—are divisible by 8 because 246000 ends with 000 and 128 of 5681009128 is divisible by 8.

9: A number is divisible by 9 if the sum of its digits is divided by 9.
Examples: 12456 is divisible by 9 because $1 + 2 + 4 + 5 + 6 = 18$, is divisible by 9.

2456791353 is divisible by 9 because $2 + 4 + 5 + 6 + 7 + 9 + 1 + 3 + 5 + 3 = 45$, is divisible by 9.

10: A number is divisible by 10 if it ends with 0.
Example: 1250, 4560, 1598750 are divisible by 10 as all of them end with 0.

11: A number is divisible by 11 if the difference between the sum of digits at even places and the sum of digits at odd places is either 0 or a multiple of 11.

Examples: 1331 is divisible by 11.
Sum of digits at odd places = $1 + 3 = 4$
Sum of digits at even places = $1 + 3 = 4$
Difference: $4 - 4 = 0$

31415 is not divisible by 11.
Sum of digits at odd places = $5 + 4 + 3 = 12$
Sum of digits at even places = $1 + 1 = 2$
Difference: $12 - 2 = 10$

There is one better rule to check the divisibility of a number by 11. Take the alternating sum of the digits in the number and read from left to right. If that is divisible by 11, so is the original number.

Example: 2728 has an alternating sum of digits $2 - 7 + 2 - 8 = -11$. Since -11 is divisible by 11, so is 2728.

Again, in the case of 31415, the alternating sum of digits is $3 - 1 + 4 - 1 + 5 = 10$. This is not divisible by 11, so neither is 31415.

13: Add four times the last digit to the remaining leading truncated number. Repeat this process until you are left with a two-digit number. If this number is divisible by 13 then the whole number is divisible by 13.

Example: Is 169 divisible by 13?
Solution: 169: $- 16 + 4 \times 9 = 52$
$5 + 4 \times 2 = 13$ is divisible by 13.
Hence, 169 is divisible by 13.
Let's do the division.

Example: Divide 1680 by 140.
Solution: Divisor is first written as the product of primes.
$140 = 2 \times 2 \times 5 \times 7$

2	1680	
5	840	0
7	168	0
2	24	0
	12	

Quotient = 12
Remainder = 0

This method is not suitable. Let's explore another interesting method. In an article published in *The Science Explorer* written by Elizabeth Knowles, a very interesting method of division was explained. There, dots have been used to find the quotient and remainder.

Here are the steps:

- Draw dots on a piece of paper in columns where each column has a number of dots that represent a digit in the number you are dividing.
- Trace lines between the dots keeping the divisor in mind.
- Repeat the process.

Example: Divide 39 by 12.
Solution: Join 3 and 9 dots separately. Join 1 – 2 dots beginning from the first dots.

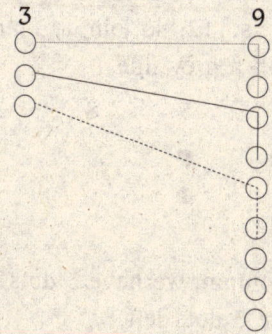

Here, we get 3 such pairs that connect 1 – 2 dots represented by grey line-black line-dotted line and 3 dots in the second column are left unattended.

Hence, Dividend = 39; Divisor = 12; Quotient = 3 (Pairs); and Remainder = 3 (Unattended dots)

Example: Divide 145824 by 112.
Solution: Here, divisor is 112 and dividend is 145824.

Write the dividend 145824 at the top and below each of the digits make as many dots as the face value of the digits represent. In simple words, below 1 put 1 dot, below 4 put 4 dots, below 5 put 5 dots, below 8 put 8 dots, below 2 put 2 dots, and below 4 put 4 dots.

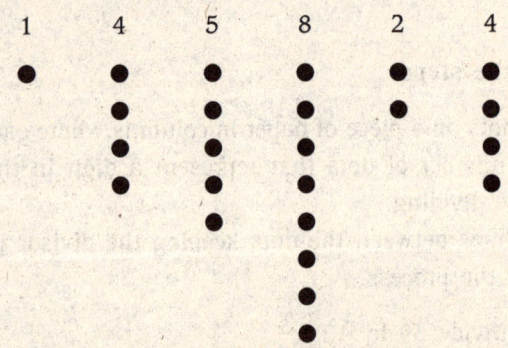

Once the dividend is represented by dots, it is time to divide the number. Join as many dots of the dividend as the value of divisor. Here, divisor is 112; so join 1, 1 and 2 dots that have been shown here with a grey line.

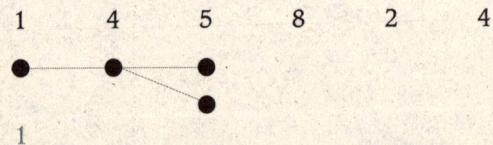

Now in the second column, we have 3 dots left and in the third column too, there are 3 dots left.

Start from the second column and join 1, 1 and 2 dots again and again and you will see that there are 3 such combinations. Hence the second digit of the quotient is 3.

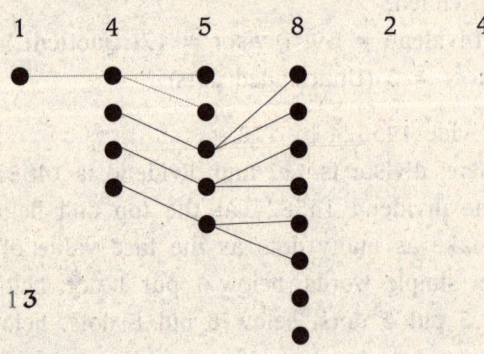

As you can see, there are 2 dots left in the fourth column but no dots left in the third column; hence the third digit of the quotient is 0. Finally, starting with the fourth column again, we join 1, 1 and 2 dots and as we can see, there are no dots left in the fifth and sixth columns. Hence division is complete with the quotient being 1302 and remainder 0.

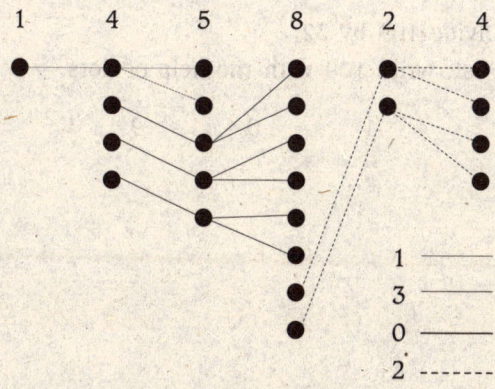

Example: Divide 133342 by 121.
Solution: First draw the dots for each digit of the number 133342 and join 1 – 2 – 1 dots.

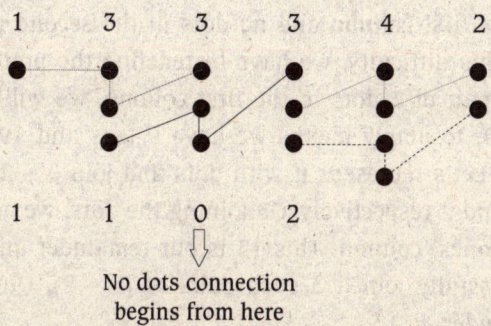

Here grey line, black line, jarred line and dotted line are used to connect the dots. The final answer, as you can see, is 1102. One line from 1 joins 1 – 2 – 1 dots, another line from 3 joins 1 – 2 – 1 dots. This process keeps going until all the dots are

The Egyptian Method of Division 83

joined. Finally, we count the dot-joining lines from each digit.

Many a times, joining dots to divide gets a little tricky when there aren't enough dots in a particular column. In such a case, you will have to transform one of the dots from a column into ten dots in the following column. Let's take an example to understand this scenario.

Example: Divide 109 by 32.
Solution: First, write 109 with the help of dots.

```
        1           0           9
        ○                       ○
        ○                       ○
        ○                       ○
                                ○
                                ○
                                ○
                                ○
                                ○
                                ○
```

Since the divisor is 32, we can't connect 3 dots from the first column with the 2 dots of the second column as there is only 1 dot in the first column and no dots in the second column. To overcome this difficulty, we have to redefine the problem. Since we need a pair of 3 dots in the first column, we will break 109 as 90 + 19. It simply means we have 9 tens and 19 ones that make 109. Let's represent it with dots and join 3 – 2 dots from column 1 and 2 respectively. On joining the dots, we are left with 13 dots in ones' column. This 13 is our remainder and since we have successfully joined 3 such pairs of (3 – 2), Quotient = 3 and Remainder = 13.

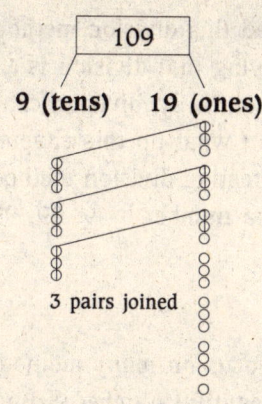

Example: Divide 3328 by 104.

Solution: Here, divisor is 104 but the third digit (from the left) of 3328 is 2, which is less than the third digit of the divisor. To complete the division, we shift 10 dots to the next column reducing 1 from the second. Here, grey dots of the second column represent 10 dots of the third column. Join 1 dot of the first column with no dots of second and 4 dots of the third column. You can see that now we have 3 black lines and 2 dotted lines joining all the dots with no dots left out at the end.

Quotient = 32
Remainder = 0

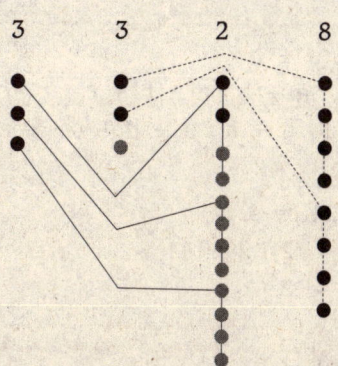

Hope you have enjoyed this division method? So the next time, if you find anyone saying that division is a tough operation, do let them know how, contrarily, division can be interesting. And this is not all! Before I wind up this chapter, let me share with you another very interesting division method that you will have fun with: Division of a number by 9, 99, 999 ...

Division by 9

In this book, I shall focus on many methods that are somehow associated with the beautiful number 9. But first, let's begin on the division by 9!

1/9 = 0.1111 ___   2/9 = 0.2222 ___   3/9 = 0.3333 ___
4/9 = 0.4444 ___   5/9 = 0.5555 ___   6/9 = 0.6666 ___
7/9 = 0.7777 ___   8/9 = 0.8888 ___

Now, let's look at the division process. Here are the steps:

- Write the digit on the extreme left as the first digit.
- Add the next digit.
- Keep adding the next digit every time to the previous sum until all the digits are taken. The last digit of the sum (extreme right) will be the remainder and the remaining digit will give you the quotient.

```
ab ÷ 9 = a / a + b
abc ÷ 9 = a / a + b / a + b + c
abcd ÷ 9 = a / a + b / a + b + c / a + b + c + d
```

Example: Divide 24 by 9.
Solution: 24 ÷ 9 = 2 / 2 + 4
            = 2 / 6

Quotient = 2
Remainder = 6

Example: Divide 43 by 9.
Solution: $43 \div 9 = 4 / 4 + 3$
$= 4 / 7$

Quotient = 4
Remainder = 7

In case the sum of the last pair is 9, increase the previous digit by 1 and subtract 9 from the last pair to get the remainder 0.

Example: Divide 54 by 9.
Solution: $54 \div 9 = 5 / 5 + 4$
$= 5 / 9$

Quotient = 5 + 1 = 6
Remainder = 0

Example: Divide 542 by 9.
Solution: $542 \div 9 = 5 / 5 + 4 / 5 + 4 + 2$
$= 5 / 9 / 11$

Quotient = 59 + 1 = 60
Remainder = 11 − 9 = 2

Example: Divide 234571 by 9.
Solution: $234571 \div 9 = 2 / 2 + 3 / 2 + 3 + 4 / 2 + 3 + 4$
$+ 5 / 2 + 3 + 4 + 5 + 7 / 2 + 3 +$
$4 + 5 + 7 + 1$
$= 2 / 5 / 9 / 14 / 21 / 22$
$= 2 / 5 / 9 + 1 / 4 + 21 / 22$
$= 2 / 5 / 10 / 6 / 1 / 22$
$= 26061 / 22$

Since the remainder 22 > 9, we divide 22 again to get the actual remainder, i.e.

$22 \div 9 = 2 / 2 + 2$
$= 2 / 4$

Therefore, actual remainder = 4 and quotient will get increased by 2.

The Egyptian Method of Division

= 26061 + 2 / 4

= 26063/ 4

Quotient = 26063

Remainder = 4

If the divisor is 99 then the same process will be applied, but this time you need to add two numbers at a time. The last two-digit number will be the remainder. In case the sum of the last pair is 99, then add 1 to the previous column and make the remainder zero. If the last pair is more than 99, repeat the division of the last pair as shown in the last example of division by 9.

Example: Divide 123134 by 99.

Solution: Here, dividend has two digits. Let's see how that works.

123134 ÷ 99 = 12 / 12 + 31 / 12 + 31 + 34

= 12 / 43 / 77

Quotient = 1243

Remainder = 77

Example: Divide 23143245 by 99.

Solution: 23143245 ÷ 99 = 23 / 23 + 14 / 23 + 14 + 32 / 23 + 14 + 32 + 45

= 23 / 37 / 69 / 114

Here, 114 > 99

114 ÷ 99 = 0114 ÷ 99

= 01 / 01 + 14

= 01 / 15

Hence, 23143245 ÷ 99 = 23 / 23 + 14 / 23 + 14 + 32 / 23 + 14 + 32 + 45

= 23 / 37 / 69 / 114

= 23 / 37 / 69 + 1 / 15

= 23 / 37 / 70 / 15

Quotient = 233770

Remainder = 15

Here, we have seen the unconventional method used in Egypt that was purely based on the binary number system used at that time in place of the decimal number system that we use today. The dot method is as brilliant and so is the division by 9, 99, etc. There is a need to explore various division methods to see how effective they can be, and that can be achieved only through practice. So keep practising and exploring!

Practice Problems

Divide each of the following using the Egyptian Method and the Dot Method:

a) 4532 ÷ 54
b) 2101532 ÷ 79
c) 12345 ÷ 15
d) 11223 ÷ 99
e) 4705 ÷ 25
f) 986048 ÷ 994
g) 4701838 ÷ 1296
h) 2448 ÷ 17
i) 1470 ÷15
j) 865 ÷173
k) 245673 ÷ 9
l) 12122055 ÷ 99

Auxiliary Fraction

You all know that there are two types of decimal fractions—terminating and non-terminating. Auxiliary fraction is a method to deal with non-terminating recurring decimals. The basic idea behind the use of auxiliary fraction is to convert the division into a modified division technique to solve division problems in an easier way without actually indulging in division.

The method is based on the following principle:

a) Divide the numerator and denominator by the appropriate power of 10.
b) Try to make the unit digit of denominator 9 if possible.

Divisors ending with 1, 6, 7, 8 and 9 can be solved using the speedy Vedic method, e.g. 21, 29, 67, etc. Other divisors ending with 2, 3, 4, 5 and 7 can be converted to such divisors by multiplying with a suitable number.

$$\frac{1}{17} = \frac{1 \times 7}{17 \times 7} = \frac{7}{119} \qquad \frac{4}{13} = \frac{4 \times 3}{13 \times 3} = \frac{12}{39}$$

The precondition to dividing numbers like that is that the dividend should be less than the divisor. Let's see how that works.

Case 1: If Divisor Ends With 9

- Add 1 to the divisor (denominator). This new denominator will act as a divisor.
- Break the denominator as a multiple of 10.
  $29 + 1 = 30 = 3 \times 10$
  $39 + 1 = 40 = 4 \times 10$
- Divide the numerator by 10 or one of its multiples* or place a decimal point in the numerator at the appropriate place.
- Now divide the dividend by the new divisor. The remainder in each case will be placed before the digit to be divided, making a new quotient each time. The division here is not an ordinary division.

Example: Divide 3 by 19.
Solution

Step 1:

- First increase the divisor by 1.
  $3/20 = 3/(2 \times 10)$
- Divide the numerator by 10.
  $3/(2 \times 10) = 0.3/2$
- Place the decimal point first and divide 3 by 2.

$$\begin{array}{r} 2)\,3\,(1 \\ \underline{-2} \\ 1 \end{array}$$

Quotient = 1
Remainder = 1
As stated above, place the remainder below the quotient.
$0.3/2 = 0._{\,1}1$
New dividend = 11

---

*Division of a number by 10 or its multiple is easier; so, the denominator is made the multiple of 10 first by adding an appropriate digit.

Auxiliary Fraction

Step 2: Divide 11 by 2.

```
2) 11 (5
   -10
   ---
    1
```

Quotient = 5
Remainder = 1
New Dividend = 15
0.3/2 = 0. $_1$1$_1$5

Step 3: Divide 15 by 2.
Quotient = 7
Remainder = 1
New Dividend = 17
Hence, 0.3/2 = 0. $_1$1 $_1$5$_1$7

Step 4: Divide 17 by 2.
Quotient = 8
Remainder = 1
New Dividend = 18
Hence, 0.3/2 = 0. $_1$1 $_1$5$_1$7$_1$8

Step 5: Divide 18 by 2.
Quotient = 9
Remainder = 0
New Dividend = 09
Hence, 0.3/2 = 0. $_1$1 $_1$5 $_1$7 $_1$8 $_0$9

Step 6: Divide 09 by 2.
Quotient = 4
Remainder = 1
New Dividend = 14
Hence, 0.3/2 = 0. $_1$1 $_1$5 $_1$7 $_1$8 $_0$9 $_1$4

Step 7: Divide 14 by 2.
Quotient = 7
Remainder = 0
New Dividend = 07

Hence, $0.3/2 = 0._1 1_1 5_1 7_1 8_0 9_1 4_0 7$

Step 8: Divide 07 by 2.
  Quotient = 3
  Remainder = 1
  New Dividend = 13
  Hence, $0.3/2 = 0._1 1_1 5_1 7_1 8_0 9_1 4_0 7_1 3$

This will be continued. Hence, the final answer, with up to 8 digits of decimal is given here.

$$\frac{3}{19} = 0.1589473\_\_\_$$

(The digit on the top is written as the final answer.)

Example: Divide 7 by 39 up to 8 decimal places.

Solution: The previous example has been explained in detail. There is no need to follow so many steps as the whole process can be performed mentally in one line. Let me show you the one-line process of division before discussing the elaborate process.

First, increase the denominator by 1, i.e. $39 + 1 = 40$

$$\frac{7}{40} = \frac{7}{4 \times 10} = \frac{0.7}{4}$$

Now do the division as discussed. Place the quotient at the top and the denominator below the respective quotient to get the new dividend each time.

Upon division we get,

$\frac{0.7}{4} = 0._3 1_3 7_1 9_3 4_2 8_0 7_3 1_3 7\_\_\_$

Hence, $\frac{7}{39} = \mathbf{0.17948717} \ldots$

Explanation:

Step 1: Divide 7 by 4.
  Quotient = 1
  Remainder = 3
  New dividend = 31
  $\frac{0.7}{4} = 0._3 1$

(Remainder is placed below the quotient each time.)

Auxiliary Fraction

Step 2: Divide 31 by 4.
   Quotient = 7
   Remainder = 3
   New dividend = 37
   $^{0.7}/_4 = 0.\ _31\ _37$
   (Remainder is placed below the quotient each time.)

Step 3: Divide 37 by 4.
   Quotient = 9
   Remainder = 1
   New dividend = 19
   $^{0.7}/_4 = 0.\ _31\ _37\ _19$
   (Remainder is placed below the quotient each time.)

Step 4: Divide 19 by 4.
   Quotient = 4
   Remainder = 3
   New dividend = 34
   $^{0.7}/_4 = 0.\ _31\ _37\ _19\ _34$
   (Remainder is placed below the quotient each time.)

Step 5: Divide 34 by 4.
   Quotient = 8
   Remainder = 2
   New dividend = 28
   $^{0.7}/_4 = 0.\ _31\ _37\ _19\ _34\ _28$
   (Remainder is placed below the quotient each time.)

Step 6: Divide 28 by 4.
   Quotient = 7
   Remainder = 0
   New Dividend = 07
   $^{0.7}/_4 = 0.\ _31\ _37\ _19\ _34\ _28\ _07$
   (Remainder is placed below the quotient each time.)

Did you understand the process?

While writing the final answer, write only the quotient.
Hence, $7/39$ = **0.17948717 ...**

Example: Divide 41 by 119.
Solution: Increase the denominator by 1.
$$41/120 = 4.1/12$$
$$= 0.\ _53\ _54\ _64\ _45\ _93\ _97\ _18\ _61 \_\_$$
Hence, $41/119$ = **0.34453781...**

Example: Divide 61 by 109.
Solution: Increase the denominator by 1.
$$61/110 = 6.1/11$$
$$= 0.\ _65\ _{10}5\ _69\ _36\ _33\ _03\ _30\ _82 \_\_$$
Hence, $61/109$ = **0.55963302...**

Practice Problems

Divide the following:

a) 4/29 b) 71/79 c) 84/139 d) 3/13
e) 1/23 f) 31/149 g) 1/17 h) 61/89
i) 78/129 j) 5/23 k) 17/69

Case 2: When Divisor Ends With 8

If divisor ends with 8, such as 18, 28, 38, 58, 88, etc., the division process, as discussed in the case of 9, will change a little bit. Let's see the method.

- Add 2 to the denominator to make it end with zero.
- Divide the numerator with 10 and place the decimal accordingly.
- Now divide the numerator with the digit at the denominator.
- The division is not an actual division, so place the remainder before the quotient at every step.

- Since the last digit of the divisor is 8, which is 1 less than 9, multiply the quotient by 1 and add to the base dividend at each step to get the gross dividend.

Example: Divide 9 by 28.
Solution: Here divisor is 28. Add 2 to make it 30.

$$9/_{30} = 9/_{3 \times 10} = 0.9 \div 3$$

Step 1: $0.9 \div 3$

Quotient = 3
Remainder = 0
$9/_{28} = 0._03$
Base dividend = 03
New dividend = Base dividend + Quotient × 1
$\qquad = 03 + 3 \times 1$
$\qquad = 06$

Step 2: $06 \div 3$

Quotient = 2
Remainder = 0
$9/_{28} = 0._03\,_02$
Base dividend = 02
New dividend = Base dividend + Quotient × 1
$\qquad = 02 + 2 \times 1$
$\qquad = 04$

Step 3: $04 \div 3$

Quotient = 1
Remainder = 1
$9/_{28} = 0._03\,_02\,_11$
Base dividend = 11
New dividend = Base dividend + Quotient × 1
$\qquad = 11 + 1 \times 1$
$\qquad = 12$

Step 4: 12 ÷ 3
   Quotient = 4
   Remainder = 0
   $9/28 = 0.\,_03\,_02\,_11\,_04$
   Base dividend = 04
   New dividend = Base dividend + Quotient × 1
         = 04 + 4 × 1
         = 08

Step 5: 08 ÷ 3
   Quotient = 2
   Remainder = 2
   $9/28 = 0.\,_03\,_02\,_11\,_04\,_22$
   Base dividend = 22
   New dividend = Base dividend + Quotient × 1
         = 22 + 2 × 1
         = 24

Step 6: 24 ÷ 3
   Quotient = 8
   Remainder = 0
   $9/28 = 0.\,_03\,_02\,_11\,_04\,_22\,_08$
   Base dividend = 08
   New dividend = Base dividend + Quotient × 1
         = 08 + 8 × 1
         = 16

Step 7: 16 ÷ 3
   Quotient = 5
   Remainder = 1
   $9/28 = 0.\,_03\,_02\,_11\,_04\,_22\,_08\,_15$
   This will continue likewise.
   Hence $9/28 = 0.3214285$

There is absolutely no need to do so many operations as everything can be done in a single line. I am taking one more example in detail and then I will show you the shorter (single-line) method.

Example: Divide 73 by 138.

Solution: Here, the divisor is 138. Add 2 to make it 140.

$$^{73}/_{140} = {}^{73}/_{14 \times 10} = 7.3 \div 14$$

Step 1: $7.3 \div 14$

Quotient = 5

Remainder = 3

$^{73}/_{138} = 0._{\,3}5$

Base dividend = 35

New dividend = Base dividend + Quotient × 1
$$= 35 + 5 \times 1$$
$$= 40$$

Step 2: $40 \div 14$

Quotient = 2

Remainder = 12

$^{73}/_{138} = 0._{\,3}5\,_{12}2$

Base dividend = 122

New dividend = Base dividend + Quotient × 1
$$= 122 + 2 \times 1$$
$$= 124$$

Step 3: $124 \div 14$

Quotient = 8

Remainder = 12

$^{73}/_{138} = 0._{\,3}5\,_{12}2\,_{12}8$

Base dividend = 128

New dividend = Base dividend + Quotient × 1
$$= 128 + 8 \times 1$$
$$= 136$$

Step 4: $136 \div 14$

Quotient = 9

Remainder = 10

$^{73}/_{138} = 0._{\,3}5\,_{12}2\,_{12}8\,_{10}9$

Base dividend = 109

New dividend = Base dividend + Quotient × 1
= 109 + 9 × 1
= 118

Step 5: 118 ÷ 14
Quotient = 8
Remainder = 6
$^{73}/_{138}$ = 0. $_3$5 $_{12}$2 $_{12}$8 $_{10}$9 $_6$8
Base dividend = 68
New dividend = Base dividend + Quotient × 1
= 68 + 8 × 1
= 76

Step 6: 76 ÷ 14
Quotient = 5
Remainder = 6
$^{73}/_{138}$ = 0. $_3$5 $_{12}$2 $_{12}$8 $_{10}$9 $_6$8 $_6$5

This division will keep going likewise. Follow the steps and find the number of digits after decimal point as per your choice. The final answer is: $^{73}/_{138}$ = 0.528985___

Now, you can do the same in a single line.
$^{73}/_{138}$ = $^{73}/_{140}$ = $^{73}/_{14 \times 10}$
= 7.3 ÷ 14
= 0. $_3$5 $_{12}$2 $_{12}$8 $_{10}$9 $_6$8 $_6$5

Practice Problems

Find the value of the following auxiliary fractions up to 7 digits after decimal:

a) 1/18   b) 3/28   c) 71/88   d) 1/48
e) 3/14   f) 43/98   g) 35/98   h) 2/38
i) 45/78   j) 17/18   k) 21/98   l) 15/78

Case 3: When the Denominator Ends with 7

As explained in the two cases above, we need to first make the denominators multiples of 10. Since denominator in this case ends with 7, we will add 3 to the denominator to make it a multiple of 10 and use this as divisor.

Here's the step-by-step explanation.

- Remove the zeros from the denominator and place a decimal in the numerator at an appropriate place.
- Divide the numerator by the number in the denominator as is done in normal division to get the first quotient and remainder. Place the remainder below the quotient.
- Since the actual denominator ends with 7, that is 2 less than 9, multiply the quotient by 2 and add to the base dividend in each case to compute the division process.

Example: Divide 73 by 137.
Solution: Add 3 to the denominator, i.e. 137 + 3 = 140.
Now, divide 73 by 140 using the method explained.

a) Remove zero from the denominator and place the decimal point in the numerator at an appropriate place.
$73/140 = 7.3/14$

b) Place the decimal point first. On dividing 73 by 14, we get,
Quotient = 5 and Remainder = 3
$^{73}/_{140} = 0._35$
The remainder is placed at the bottom of the quotient.
Base dividend = 35
Actual dividend = Base dividend + 2 × Quotient
$= 35 + 2 \times 5$
$= 45$

c) Divide 45 by 14.
Quotient = 3
Remainder = 3
$^{73}/_{140} = 0._35\,_33$

Remainder is placed at the bottom of the quotient.
Base dividend = 33
Actual dividend = Base dividend + 2 × Quotient
  = 33 + 2 × 3
  = 39

d) Divide 39 by 14.
Quotient = 2
Remainder = 11
$^{73}/_{140}$ = 0. $_3$5 $_3$3 $_{11}$2
Base dividend = 112
Actual dividend = Base dividend + 2 × Quotient
  = 112 + 2 × 2
  = 116

e) Divide 116 by 14.
Quotient = 8
Remainder = 4
$^{73}/_{140}$ = 0. $_3$5 $_3$3 $_{11}$2 $_4$8
Base dividend = 48
Actual dividend = Base dividend + 2 × Quotient
  = 48 + 2 × 8
  = 64

f) Divide 64 by 14.
Quotient = 4
Remainder = 8
$^{73}/_{140}$ = 0. $_3$5 $_3$3 $_{11}$2 $_4$8 $_8$4
Base dividend = 84
Actual dividend = Base dividend + 2 × Quotient
  = 84 + 2 × 4
  = 92

g) Divide 92 by 14.
Quotient = 6
Remainder = 8
$^{73}/_{140}$ = 0. $_3$5 $_3$3 $_{11}$2 $_4$8 $_8$4 $_8$6
Base dividend = 86

You can continue the process if you wish. Here, we have completed the six steps of division after decimal, so we don't need to divide any more. The final answer is:

$73/137 = 0.532846\_\_\_$

Example: Divide 45 by 127.

Solution: Add 3 to the denominator, i.e. 127 + 3 = 130.

Now, divide 45 by 130 using the method explained.

a) Remove zero from the denominator and place the decimal point in the numerator at an appropriate place.

$45/130 = 4.5/13$

b) Place the decimal point first. On dividing 45 by 13, we get

Quotient = 3 and Remainder = 6

$45/13 = 0._{6}3$

The Remainder is placed at the bottom of the quotient.

Base dividend = 63

Actual dividend = Base dividend + 2 × Quotient
= 63 + 2 × 3
= 69

c) 69 ÷ 13

Quotient = 5

Remainder = 4

$45/13 = 0._{6}3\,_{4}5$

Base dividend = 45

Actual dividend = Base dividend + 2 × Quotient
= 45 + 2 × 5
= 55

d) 55 ÷ 13

Quotient = 4

Remainder = 3

$45/13 = 0._{6}3\,_{4}5\,_{3}4$

Base dividend = 34

Actual dividend = Base dividend + 2 × Quotient

$$= 34 + 2 \times 4$$
$$= 42$$

e) $42 \div 13$
   Quotient = 3
   Remainder = 3
   $^{45}/_{13} = 0.\,_{6}3\,_{4}5\,_{3}4\,_{3}3$
   Base dividend = 33
   Actual dividend = Base dividend + 2 × Quotient
   $$= 33 + 2 \times 3$$
   $$= 39$$

f) $39 \div 13$
   Quotient = 3
   Remainder = 0
   $^{45}/_{13} = 0.\,_{6}3\,_{4}5\,_{3}4\,_{3}3\,_{0}3$
   Base dividend = 03

Since we have completed the steps of division as above, it is left for you to get the next quotient. Hence, we have $^{45}/_{13} = 0.35433\_\_\_$

I do hope you have understood the concept of auxiliary fraction. So far we have seen three cases:

a) When the denominator ends with 9
b) When the denominator ends with 8
c) When the denominator ends with 7

Now the question is, will this method be applicable if the denominator ends with 6?

The answer is YES.

- When the denominator was 8, we changed the base dividend into actual dividend by applying the formula: **Actual dividend = Base dividend + 1 × Quotient (8 is 1 [9 – 8] less than 9)**
- When the denominator was 7, we changed the base dividend into actual dividend by applying the formula:

Actual dividend = Base dividend + 2 × Quotient
(7 is 2 [9 – 7] less than 9)

Similarly, in case of a denominator ending with 6, we will use—
Actual dividend = Base dividend + 3 × Quotient (6 is 3 [9 – 6] less than 9)

Hope, the modus operandi is clear to you. Keep practising more and more problems of auxiliary fraction to achieve mastery over it. You need not write each and every detailed step as explained here; on the contrary, you can skip all the steps and do the division in your head.

Practice Problems

a) 1/17   b) 3/26   c) 71/87   d) 11/47
e) 13/36   f) 43/97   g) 35/96   h) 21/37
i) 45/76   j) 17/46   k) 21/97   l) 15/76

Division of Polynomials

The division of a polynomial by another polynomial sometimes puts an unnecessary amount of burden on our minds. But this method can be simplified with the help of 'synthesis division'. Even dividing numbers in Arithmetic in Vedic mathematics is based on synthesis division.

The main point that needs to be remembered is that the divisor should be a linear expression and its leading coefficient should be 1. You can't use $x^2 + 4$ or $2x^2 - 5x + 2$ as a divisor. Moreover, if your divisor is $4x - 7$ then it should be made $x - 7/4$ to get the coefficient of the first term as a linear expression.

Working method:

- First, put the denominator equal to zero. Put it in the first column.
- The dividend should be written in descending order of powers. In case any term is missing, put a zero to fill the missing term. In the dividend column, put only the coefficient of polynomials, not the whole term.
- Carry down the first number of dividend.
- Multiply the number placed in the divisor column by the first leading coefficient carried down and put the result in the next column.
- Add the two numbers together and write the result in the bottom.
- Repeat the same steps till you reach the end of the problem.

- The bottom row gives you the answer. The last number of the bottom row is the remainder and the remaining is the quotient of the polynomial. Starting from the left, put the power of variable as one power less than the original power and go down one with each term.

Example: Divide $2x^3 - 5x^2 + 3x + 7$ by $x - 2$

Solution:

a) As stated above, first put the denominator equal to zero.
   If $x - 2 = 0$
   $\Rightarrow x = 2$ (This is the divisor.)

b) Next, write the coefficient of the polynomial that is to be divided. The polynomial should be arranged in descending order and in case any term is missing, use zero in that place against the missing term.

$$\underline{2\,|2\quad -5\quad 3\quad 7}$$

c) Carry the first number straight down.

$$\begin{array}{r|rrrr} 2 & 2 & -5 & 3 & 7 \\ & \downarrow & & & \\ \hline & 2 & & & \end{array}$$

d) Multiply the number carried down with the divisor and write the result in the next column.

$$\begin{array}{r|rrrr} 2 & 2 & -5 & 3 & 7 \\ & \downarrow & 4 & & \\ \hline & 2 & & & \end{array}$$

e) Add the number in the second column and write the result in the bottom.

$$\begin{array}{r|rrrr} 2 & 2 & -5 & 3 & 7 \\ & \downarrow & 4 & & \\ \hline & 2 & -1 & & \end{array}$$

f) Multiply the second number in the column by the divisor and write it below the next column.

$$\begin{array}{r|rrrr} 2 & 2 & -5 & 3 & 7 \\ & \downarrow & 4 & -2 & \\ \hline & 2 & -1 & & \end{array}$$

g) Now add the two numbers of the third column and write the sum below.

$$\begin{array}{r|rrrr} 2 & 2 & -5 & 3 & 7 \\ & \downarrow & 4 & -2 & \\ \hline & 2 & -1 & 1 & \end{array}$$

h) Multiply the third number in the bottom line with the divisor and write it below the fourth column.

$$\begin{array}{r|rrrr} 2 & 2 & -5 & 3 & 7 \\ & \downarrow & 4 & -2 & 2 \\ \hline & 2 & -1 & 1 & \end{array}$$

i) Add the two numbers in the fourth column.

$$\begin{array}{r|rrrr} 2 & 2 & -5 & 3 & 7 \\ & \downarrow & 4 & -2 & 2 \\ \hline & 2 & -1 & 1 & ⑨ \end{array}$$

The division is complete. The last digit of the bottom line which is encircled is the remainder. Now you can write the result down. The bottom line gives the quotient and remainder together. In order to write the quotient, diminish one power of the polynomial and place it with each term from the leftmost digit to the right. The power of the number should also be written in the descending order.

Dividend = $2x^3 - 5x^2 + 3x + 7$
Divisor = $x - 2$
Quotient = $2x^2 - x + 1$
Remainder = 9

The whole operation can be done in a single line. The best thing about this method is its simplicity. No worrying with subtraction, tough multiplications and the arrangement of power. Here's another example.

Example: Divide $x^3 + 6x^2 + 11x + 6$ by $x + 3$.

Solution: First, put $x + 3$ equal to zero to get the digit to be placed in the divisor column.

$\Rightarrow x + 3 = 0$

$\Rightarrow x = -3$

Place the coefficients of the polynomial—1, 6, 11, 6 in the dividend column.

$$\begin{array}{c|cccc} -3 & 1 & 6 & 11 & 6 \\ & \downarrow & & & \\ \hline \end{array}$$

Now carry the first digit down. Multiply it by the divisor and place it in the next column.

$1 \times -3 = -3$

Add the second column.

$6 + (-3) = 3$

Multiply the digit placed at the bottom of the second column by the divisor and place it below the third column.

$3 \times -3 = -9$

Add the digits in the third column.

$$\begin{array}{c|cccc} -3 & 1 & 6 & 11 & 6 \\ & \downarrow & -3 & -9 & -6 \\ \hline & 1 & 3 & 2 & | \; 0 \end{array}$$

Finally, multiply the third column sum by the divisor and place it below the fourth column.

$2 \times -3 = -6$

Add

$6 + (-6) = 0$

Hence, Quotient = $x^2 + 3x + 2$
Remainder = 0

Example: Divide $2x^3 + 5x^2 + 9$ by $x + 3$.
Solution: Put $x + 3 = 0$
$\Rightarrow x = -3$

Arrange the dividend in descending order of power, i.e.

$2x^3 + 5x^2 + 0x + 9$

Now place the coefficient of dividend and divisor as shown below—

$$\underline{-3}\ |\ 2 \quad 5 \quad 0 \quad 9$$

Carry down the first digit (2), multiply it by divisor (–3) and place the result ($2 \times -3 = -6$) in the next column.

$$\begin{array}{r|rrrr} -3 & 2 & 5 & 0 & 9 \\ & \downarrow & -6 & & \\ \hline & 2 & & & \end{array}$$

Add the digit in the second column and write the result ($5 - 6 = -1$) below. Multiply it again with the divisor and place the result ($-1 \times -3 = 3$) below the third column.

Add the two digits of the third column and place the sum at the bottom.

$$\begin{array}{r|rrrr} -3 & 2 & 5 & 0 & 9 \\ & \downarrow & -6 & 3 & \\ \hline & 2 & -1 & 3 & \end{array}$$

Finally, multiply the result obtained at the third column by the divisor ($3 \times -3 = -9$).

$$\begin{array}{r|rrrr} -3 & 2 & 5 & 0 & 9 \\ & \downarrow & -6 & 3 & -9 \\ \hline & 2 & -1 & 3 & 0 \end{array}$$

Hence, for the polynomial
$p(x) = 2x^3 + 5x^2 + 9$
Quotient = $2x^2 - x + 3$
Remainder = 0

Let's take one more example.

Example: Divide $x^3 + 1 = 0$ by $x + 1$
Solution: Here, polynomial $p(x) = x^3 + 1$

First write the polynomial in descending order of their power.
$p(x) = x^3 + 0x^2 + 0x + 1$
Now, put $x + 1 = 0$
$\Rightarrow x = -1$

Now make a division box and place the coefficient of polynomials in the dividend part and put –1 in the divisor part.

The above process of synthetic division is a cake walk and you need not write every step down if you have understood the method clearly.

$$\begin{array}{r|rrrr} -1 & 1 & 0 & 0 & 1 \\ & \downarrow & -1 & 1 & -1 \\ \hline & 1 & -1 & 1 & 0 \end{array}$$

The highest power in $x^3 + 1$ is 3, so while writing the quotient, start with power 2 and keep decreasing with 1 in the subsequent stage.

Hence, quotient = $x^2 - x + 1$
Remainder = 0

Practice Problems

a) Divide $x^3 + 6x^2 + 11x + 6$ by $x + 3$
b) Divide $x^3 - 10x^2 - 53x - 42$ by $x - 7$
c) Divide $x^3 - 3x^2 - 9x - 5$ by $x - 1$
d) Divide $x^3 - 1$ by $x - 1$
e) Divide $x^3 + 13x^2 + 31x - 45$ by $x + 5$

IV
FRACTIONS

As Easy As Pie

You must have heard the story I am about to narrate: A man died, leaving a will. He had 17 camels that he wanted to give to his three sons in such a way that the eldest son got $1/2$, the second son got $1/3$ and the third son got $1/9$. Dividing the camels according to the will was not possible so the sons finally decided to go to the king. The king heard their problem and gave them one camel upon the condition that they will return the camel once the division of the total number of camels amongst the brothers was complete. All the sons returned home taking one camel from the king, divided the other camels in accordance to the will and then returned the one camel given to them by the king.

Total number of camels = 17
Camel given by the king = 01
New total = 18
Share of eldest son = $1/2$ of the total camels = 9 camels
Share of second son = $1/3$ of the total camels = 6 camels
Share of third son = $1/9$ of total camels = 2 camels
Remaining camel that was finally returned to the king = 01

No one knows the truth of this story but this tells us the importance of fraction in one's life. I shall not focus on the basics of fraction that are taught at the elementary level but on the contrary, focus on the techniques which can be used to lessen the time required for calculations.

Addition of Fractions

Addition of fractions is simple and you have done it in the upper primary level itself. Here, I shall take some cases which are common and generally asked in the competitive examinations, as examples.

Case 1: When the two fractions to add have a numerator 1

Shortcut Formula: Sum of the denominator/ Product of the denominator

Example: 1/5 + 1/6
Solution: Here the denominators are 5 and 6.
  Sum of the denominators, 5 + 6 = 11
  Product of the denominators, 5 × 6 = 30
  Hence, 1/5 + 1/6 = 11/30

Example: 1/7 + 1/9
Solution: Here the denominators are 7 and 9.
  Sum of the denominators, 7 + 9 = 16
  Product of the denominators, 7 × 9 = 63
  Hence, 1/7 + 1/9 = 16/63

Case 2: When two fractions have different denominators

This is a little problematic as you have to convert each fraction into fractions that will have the same denominators. But a simple mental emphasis can help you to get the answer instantly. This can be easily done by vertical[*] and crosswise multiplication techniques.

Here are some examples employing the crosswise multiplication technique.

---

[*]Vertical addition/subtraction is valid only if the denominators are the same. It is being taught at the primary levels and doesn't come under the speed-maths techniques. Hence, I have not provided any examples related to that here.

Example: 5/7 + 4/5 = ?
Solution:

$$\frac{5}{7} + \frac{4}{5}$$

$$= \frac{25 + 28}{35} = \frac{53}{35}$$

Example: 2/3 + 5/8 = ?
Solution:

$$\frac{2}{3} + \frac{5}{8}$$

$$= \frac{16(2 \times 8) + 15(3 \times 5)}{24} = \frac{31}{24}$$

Example: 7/11 + 2/9 = ?
Solution:

$$\frac{7}{11} + \frac{2}{9}$$

$$= \frac{63(7 \times 9) + 22(2 \times 11)}{99} = \frac{85}{99}$$

Case 3: Adding two fractions when the denominators have common factors

Using the traditional method to find the answer of fractions when their denominators have common factors is a tough job; there, you will first have to find the LCM of the denominators and convert both the fractions into ones with a common denominator and follow the previous method. But there is another way out.

- Take the HCF of the denominator.
- Divide the denominator with the HCF and write the quotient obtained down below the corresponding denominator.
- For the numerator, cross-multiply the numerator with the quotient written below.

- For the denominator, cross-multiply each denominator with each quotient placed diagonally.

Example: $\frac{5}{18}+\frac{4}{27}$

Solution:

$$\frac{5}{18}+\frac{4}{27}$$

$18 = 2 \times 9$ and $27 = 3\times9$

- Take the HCF of the denominator, which is 9, and write the quotient below the denominator as shown here.

$$\begin{matrix}\frac{5}{18} & + & \frac{4}{27} \\ (2) & & (3)\end{matrix}$$

- Cross-multiply the numerator with the quotient as shown and write it as numerator.

$$\begin{matrix}\frac{5}{18} & + & \frac{4}{27} \\ (2) & & (3)\end{matrix}$$

Numerator $= 5\times3 + 4\times2 = 15 + 8 = 23$

- For denominator, cross-multiply each of the denominators with each of the quotients.

$$\begin{matrix}\frac{5}{18} & + & \frac{4}{27} \\ (2) & & (3)\end{matrix}$$

Denominator $= 18 \times 3 = 54$ or $27 \times 2 = 54$

Hence, $\frac{5}{18}+\frac{4}{27}=\frac{23}{54}$

Case 4: Adding more than two fractions

Addition of more than two fractions looks tough but it will become easy when you follow the given steps.

- Choose the fraction for which you want the operation. Circle its numerator.
- Strike the number vertically and horizontally along with the number, except itself.
- Now multiply the rest of the left-out number.
- Do the same operation for the rest of the fraction.
- For the denominator, multiply the denominator of all the fractions.

Example: $3/4 + 2/3 + 2/5$
Solution:

$$\frac{3}{4} + \frac{2}{3} + \frac{2}{5}$$

$$\cancel{3}\ \cancel{2}\ \cancel{2}\quad 3\ \cancel{2}\ 2\ \cancel{3}\ \cancel{2}\ \cancel{2}$$
$$4\ \cancel{3}\ \cancel{5}\quad 4\ \cancel{3}\ 5\ 4\ 3\ \cancel{5}$$

The pictorial representation above shows that for the first fraction, all the numbers in the horizontal line, except the numerator of the first fraction, have been struck off and the vertical numbers below the concerned numerator have also been struck off. The same is followed for the rest of the fraction.

For the first fraction whose numerator is 3, the number along it for the other fractions like 2 and 2, are cut out. Moreover, the numbers on the vertical line along with the numerator 3 is also struck off. Now the left-out numbers for the first fraction are 3, 3 and 5. Multiply these numbers and that will give you the final result for the first fraction. The second and third fractions will be treated in the same manner.

Hence the result.

$$= \frac{45 + 40 + 24}{60} = \frac{109}{60}$$

Example: $5\ 1/3 + 3\ 1/4 + 7\ 1/2$
Solution: First we convert the mixed fraction. The problem will now look like:

$$^{16}/_3 + {}^{13}/_4 + {}^{15}/_2$$

⑯ 13̶ 15̶   16̶ ⑬ 15̶   16̶ 13̶ ⑮

3   4   2   3   4   2   3   4   2

$16 \times 4 \times 2 = 128$ $13 \times 3 \times 2 = 78$ $15 \times 3 \times 4 = 180$

Hence, Numerator: $128 + 78 + 180 = 386$

Denominator: $3 \times 4 \times 2 = 24$

Fraction $= 386 / 24$

Case 5: Adding two mixed fractions

Rule: Don't break the mixed fractions into improper fractions and add them separately. Instead, add the whole parts first and fractional parts separately. If you find that the sum of the fractional parts can be further converted into mixed fractions, do it.

Example: Add $2\ ^3/_7 + 4\ ^5/_7$

Solution: Add the whole parts first: $2 + 4 = 6$

Add the fractional parts: $3/7 + 5/7 = 8/7 = 1\ ^1/_7$

Hence, $2\ ^3/_7 + 4\ ^5/_7 = 6 + 1\ ^1/_7 = 7\ ^1/_7$

Example: Add $4\ ^1/_2 + 3\ ^7/_9$

Solution: Add the whole parts first: $4 + 3 = 7$

Add the fractional parts: $^1/_2 + {}^7/_9 = \dfrac{1 \times 9 + 2 \times 7}{2 \times 9} = \dfrac{23}{18} = 1\ ^5/_{18}$

Hence, $4\ ^1/_2 + 3\ ^7/_9 = 7 + 1\ ^5/_{18} = 8\ ^5/_{18}$

Subtraction of Fractions

The subtraction of fractions is also the same and if you are proficient in handling the addition of fractions, you can do the subtraction too, in the same effective way.

Case 1: When the two fractions to be subtracted have numerator 1.

Shortcut Formula: $\dfrac{\text{Difference of the denominator}}{\text{Product of the denominator}}$

Example: $1/5 - 1/6$
Solution: Here the denominators are 5 and 6.
 Sum of denominators: $6 - 5 = 1$
 Product of the denominators: $5 \times 6 = 30$
 Hence, $1/5 - 1/6 = 1/30$

Example: $1/7 - 1/9$
Solution: Here the denominators are 7 and 9.
 Sum of denominators: $9 - 7 = 2$
 Product of the denominators: $7 \times 9 = 63$
 Hence, $1/7 - 1/9 = 2/63$

Case 2: When the denominators are different

Subtraction of fractions can be done quickly using the cross-multiplication method. The important point to remember is that while finding the numerator, the first number in subtraction will be the product of the numerator of the first fraction and the denominator of the second fraction.

Example: Solve $2/5 - 1/4$
Solution:

$$\dfrac{2}{5} \diagdown\!\!\!\!\!\diagup \dfrac{1}{4}$$

$$= \dfrac{8-5}{20} = \dfrac{3}{20}$$

Example: Solve $11/7 - 2/5$
Solution: The numerator can be found by cross-multiplication and the denominator can be found by multiplying the denominators

As Easy As Pie

of each fraction.

$$\frac{11}{7} \diagup \frac{2}{5}$$

$$\frac{11 \times 5 - 2 \times 7}{7 \times 5} = \frac{55 - 14}{35} = \frac{41}{35}$$

Example: Solve $5/9 - 3/4$

Solution: The subtraction of this fraction can be done easily in the same fashion. As said earlier, in case of subtraction we have to be a little cautious and first write the product of the numerator of the first fraction and the denominator of the second fraction.

$$\frac{5}{9} \diagup \frac{3}{4} = \frac{20 - 27}{36} = \frac{-7}{36}$$

Case 3: Subtracting two mixed fractions

In case of a mixed fraction, the operation on the whole parts and fractional parts are done separately.

Example: Solve $3\,2/7 - 1\,1/5$

Solution: First do the operation with the whole parts: $3 - 1 = 2$

Now apply subtraction on the fractional parts as shown above.

$$\frac{2}{7} - \frac{1}{5} = \frac{10 - 7}{35} = \frac{3}{35}$$

Hence, $3\,2/7 - 1\,1/5 = 2\,3/35$

Example: Solve $5\,2/5 - 2\,4/7$

Solution: Subtract the whole parts first: $5 - 2 = 3$

Now it is the time to do the operation on the fractional parts.

$$\frac{2}{5} \diagup \frac{4}{7} = \frac{14 - 20}{35} = \frac{-6}{35}$$

Here, the fractional part is negative. But no worries! You can easily deal with this. Subtract the negative part from 1, which will diminish the whole part by 1.

New fractional part = $1 - 6/35 = 29/35$
$5\ 2/5 - 2\ 4/7 = 2\ 29/35$

Case 4: Subtraction from the whole number

When you subtract a fraction from a whole number, first divide the whole part into two parts, making the second part a fraction whose value is equivalent to 1. Finally, subtract the fractional part from the second part of the whole that is made into a fraction.

Example: $12 - 3/7$
Solution: Here the whole part is 12.
$12 = 11 + 1$
$ = 11 + 7/7$
Now, $12 - 3/7 = 11 + 7/7 - 3/7$ [†]
$ = 11 + 4/7$

Example: $17 - 4\ 3/11$
Solution: $17 - 4\ 3/11$
$= 17 - 4 - 3/11$
$= 13 - 3/11$
$= 12 + 11/11 - 3/11$
$= 12\ 8/11$

Example: $27 - 7\ 1/7$
Solution: Write the mixed fraction $7\ 1/7$ as $7 + 1/7$
Hence, $27 - 7\ 1/7 = 27 - 7 - 1/7$
$ = 20 - 1/7$
$ = 19 + 1 - 1/7$
$ = 19 + 7/7 - 1/7$
$ = 19\ 6/7$

[†] Since the denominator is same in this case, the vertical method can be used here. Simply subtract the numerator and there is no change in the denominator.

As Easy As Pie

Case 5: Subtraction when the fraction is mixed

Subtraction of mixed fractions can be done without changing the mixed fractions into improper fractions. Here are the steps that you need to follow:

1. Write the mixed fraction in two columns. In one column, place the whole parts; under another, the fractional parts.
2. Make the fractional parts have an equal denominator.
3. Subtract the whole part from the whole and fractional part from the fractional part.
4. In case the result of the fractional part comes out to be negative, please subtract from the whole part as done in the previous method.

Example: $4\ ^3/_4 - 1\ ^3/_7$
Solution:

$$
\begin{array}{r|l}
4 & 3/4 = \dfrac{3 \times 7}{4 \times 7} \\
-1 & 3/7 = \dfrac{3 \times 4}{7 \times 4} \\
\hline
3 & 21-12 \\
 & \overline{28}
\end{array}
$$

Hence, $-4\ ^3/_4 - 1\ ^5/_7 = 3\ ^9/_{28}$

Example: $7\ ^2/_9 - 1\ ^3/_8$
Solution:

$$
\begin{array}{r|l}
7 & 2/9 \\
1 & 3/8
\end{array}
\qquad
\begin{array}{r|l}
7 & \dfrac{2 \times 8}{9 \times 8} \\
-1 & \dfrac{3 \times 9}{8 \times 9} \\
\hline
6 & 16-27 \\
 & \overline{72}
\end{array}
$$

Now, $6 + \dfrac{(-11)}{72} = 5 + 1 - \dfrac{11}{72} = 5 + \dfrac{72}{72} - \dfrac{11}{72}$

$$= 5\ \dfrac{61}{72}$$

Multiplication of Fractions

Multiplication of two fractions or more, in case they all are proper or improper, is easy. In such cases, simply multiply the numerator and denominator separately after cancelling out the terms which can be cancelled. Write the final answer by reducing the fraction into the smallest factor.

Example: Solve $\dfrac{2}{15} \times \dfrac{5}{12} \times \dfrac{10}{70}$

Solution:

$$\dfrac{{}^1\cancel{2}}{\cancel{15}_3} \times \dfrac{{}^1\cancel{5}}{\cancel{12}_6} \times \dfrac{\cancel{10}}{\cancel{70}} = \dfrac{1}{126}$$

Example: Solve $\dfrac{12}{21} \times \dfrac{8}{60} \times \dfrac{7}{5}$

Solution:

$$\dfrac{{}^1\cancel{12}}{\cancel{21}_3} \times \dfrac{\cancel{8}}{\cancel{60}_5} \times \dfrac{\cancel{7}^1}{\cancel{5}} = \dfrac{8}{75}$$

Simplifying fractions can be easily done if you know the divisibility rule.[†]

To multiply a number by $^3/_4$

Rule (a): Divide the number by 4 and subtract it from the number itself.

Explanation: $1 - {}^1/_4 = 3/4$

Example: $24 \times {}^3/_4$
Solution: $24 \times {}^3/_4 = 24 - {}^1/_4 \times 24$
$= 24 - 6 = 18$

[†]Please see Section III of this book for the explanation of the divisibility rule.

Example: $156 \times {}^3/_4$
Solution: $156 \times {}^3/_4$
$= 156 - {}^1/_4 \times 156$
$= 156 - 39 = 117$

Example: $1728 \times {}^3/_4$
Solution: $1728 - 1728 \times {}^1/_4$
$= 1728 - 432$
$= 1296$

Example: $129 \times {}^3/_4$
Solution: $129 \times {}^3/_4 = 129 - {}^1/_4 \times 129$
$= 129 - 32.25$
$= 96.75$

Let's do some more problems for practice.

Example: Solve $128 \times {}^3/_4$
Solution: Halve the number two times successively and add the result.

$128 \div 2 = 64$ and $64 \div 2 = 32$
Add the results: $64 + 32 = 96$
Hence, $128 \times {}^3/_4 = 96$

Example: $67 \times {}^3/_4$
Solution: $67 \times {}^1/_2 = 33.5$
Hence, $67 \times {}^3/_4 = 67 \times {}^1/_2 + {}^1/_2 \times 33.5$
$= 33.5 + 16.25$
$= 49.75$

To multiply a number by $2\,{}^1/_2$

If you are asked to multiply a number by $2\,{}^1/_2$ your first reaction will be to convert $2\,{}^1/_2$ into a fraction and then solve it.

In multiplication, we have learnt how to multiply a number by 10.

So, $10 \div 4 = 2\,{}^1/_2$

I shall use the same concept here. Divide the number by 4 and put one zero at the end of the result obtained. You can also solve it in another way which would just be a simplification of the method explained. Divide the number by 2 two times and put a zero at the end of the final result.

Example: Multiply 12 by $2\frac{1}{2}$.
Solution: Divide 12 by 4.
$12 \div 4 = 3$
Now put a zero at the end of the result obtained.
Hence, $12 \times 2\frac{1}{2} = 30$.

Example: Multiply 546 by $2\frac{1}{2}$.
Solution: First put a zero after the number to be multiplied. This will make 546 as 5460. Now Divide 5460 by 4.
$5460 \div 4 = 1365$
Hence, $546 \times 2\frac{1}{2} = 1365$

Example: Multiply 2468 by $2\frac{1}{2}$.
Solution: Divide 2468 by 4.
$2468 \div 4 = 617$
Now put a zero at the end of the result obtained.
Hence, $2468 \times 2\frac{1}{2} = 6170$

Multiply two mixed numbers when they both end in ½ and the sum of their whole numbers is even

Rule: a) Multiply the whole numbers.
b) Take average of the whole numbers or divide the sum of the whole numbers by 2.
c) Add the result and place $\frac{1}{4}$ at the end.

Example: $8\frac{1}{2} \times 4\frac{1}{2}$
Solution: Multiply the whole parts: $8 \times 4 = 32$
Take average of 8 and 4: $(8 + 4) \div 2 = 6$
Add both the results: $32 + 6 = 38$

Put ¼ with the previous result: 38 $\frac{1}{4}$
Hence, 8 $\frac{1}{2}$ × 4 $\frac{1}{2}$ = 38 $\frac{1}{4}$

Example: 12 $\frac{1}{2}$ × 8 $\frac{1}{2}$
Solution: Multiply the whole parts: 12 × 8 = 96
Take average of 12 and 8: (12 + 8) ÷ 2 = 10
Add both the results: 96 + 10 = 106
Attach $\frac{1}{4}$ to the previous result: 106 $\frac{3}{4}$
Hence, 12 $\frac{1}{2}$ × 8 $\frac{1}{2}$ = 106 $\frac{3}{4}$

Multiply two mixed numbers when both end in ½ and the sum of their whole numbers is odd

Rule: a) Multiply the whole number.
b) Take the average of the whole numbers or divide the sum of whole numbers by 2 and drop the fractional parts.
c) Add both results and place $\frac{3}{4}$ at the end.

Example: 7 $\frac{1}{2}$ × 4 $\frac{1}{2}$
Solution: Multiply the whole parts: 7 × 4 = 28
Take average of 7 and 4: (7 + 4) ÷ 2 = 5 $\frac{1}{2}$
Add both the results by dropping the fractional parts: 28 + 5 = 33
Place $\frac{3}{4}$ against the previous result: 33 $\frac{3}{4}$
Hence, 7 $\frac{1}{2}$ × 4 $\frac{1}{2}$ = 33 $\frac{3}{4}$

Example: 13 $\frac{1}{2}$ × 8 $\frac{1}{2}$
Solution: Multiply the whole parts: 13 × 8 = 104
Take average of 13 and 8: (13 + 8) ÷ 2 = 10 $\frac{1}{2}$
Add both the results by dropping the fractional parts: 104 + 10 = 114
Place $\frac{3}{4}$ against the previous result: 114 $\frac{3}{4}$
Hence, 12 $\frac{1}{2}$ × 8 $\frac{1}{2}$ = 106 $\frac{3}{4}$

To multiply two mixed fractions having the same whole parts and the sum of their fractional parts is 1.

The fractional parts should sum up to 1. They should be any of these combinations or any other.

$1/2 + 1/2 = 1$
$1/4 + 3/4$
$1/3 + 2/3$
$1/5 + 4/5$

Rule: a) Multiply the whole number by its next digit.
b) Multiply the fractional parts.
c) Add the two to get the final result.

Example: Multiply $13\ 1/2$ by $13\ 1/2$.
Solution: Multiply 13 by its next digit: $13 \times 14 = 182$
Multiply the fractional parts: $1/2 \times 1/2 = 1/4$
Add the two results: $182\ 1/4$
Hence, $13\ 1/2 \times 13\ 1/2 = 182\ 1/4$

Example: Multiply $18\ 1/4$ by $18\ 3/4$.
Solution: Multiply 18 by its next digit: $18 \times 19 = 342$
Multiply the fractional parts: $1/4 \times 3/4 = 3/16$
Add the two results: $342\ 3/16$
Hence, $18\ 1/4 \times 18\ 3/4 = 342\ 3/16$

Multiplying an even number by a mixed fraction ending in ½

Method: Multiply the mixed fraction by 2 and halve the number to be multiplied.

Example: Multiply 12 by $2\ 1/2$.
Solution: $12 \times 2\ 1/2 = 12/2 \times 2 \times 2\ 1/2$
$= 6 \times 5 = 30$

Example: Multiply 24 by $3\,^1/_2$.
Solution: $24 \times 3\,^1/_2 = 24/2 \times 2 \times 3\,^1/_2$
$= 12 \times 7 = 84$

Example: Multiply 2876 by $7\,^1/_2$.
Solution: $2876 \times 7\,^1/_2 = 2876/2 \times 2 \times 7\,^1/_2$
$= 1438 \times 15$
$= 1438 \times 10 + 1438 \times ^1/_2 \times 10$
$= 14380 + 7190$
$= 21570$

Multiply two mixed fractions

Multiplying two mixed fractions involves a number of steps. We simply convert the mixed fractions into improper fractions and reduce them into the smallest fractions. But this is not the only way out. What if you could learn another way out of this odd situation? Here we shall learn the technique to multiply two mixed fractions without converting it into improper fractions.

The whole process involves the use of distributive property.
$A \times (B + C) = A \times B + A \times C$

A few years ago, I was going through an article which had used the mnemonics 'FOIL' to solve such multiplications. 'FOIL' stands for First, Outer, Inner and Last term. Let's apply this to the multiplication of mixed fractions.

Example: Multiply $5\,^1/_2$ by $4\,^1/_4$.
Solution: $5\,^1/_2$ can be written as $5 + ^1/_2$ and $4\,^1/_4$ can be written as $4 + ^1/_4$. Let's do the multiplication.
$(5 + ^1/_2) \times (4 + ^1/_4) = 5 \times 4 + 5 \times ^1/_4 + 4 \times ^1/_2 + ^1/_2 \times ^1/_4$
$= 20 + ^5/_4 + 2 + ^1/_8$
$= 20 + 1\,^1/_4 + 2 + ^1/_8$
$= (20 + 1 + 2) + (^1/_4 + ^1/_8)$
$= 23 + 3/8 = 23\,^3/_8$

But as you can see, the method above involves a number of steps. This can be solved easily if you apply your mind well and convert the fractional parts into decimal fractions.

$(5 + 1/2) \times (4 + 1/4) = 5.50 \times 4.25$

Now, can you do the multiplication by the dot and stick method? We first do the multiplication and place the decimal after the result is obtained.

Example: Multiply 550 by 425.
Solution:
Arrange the numbers on the dots as shown below.

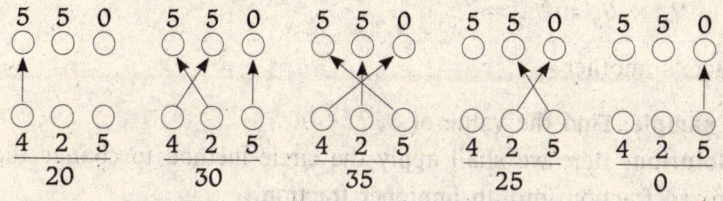

$= 20 / 30 / 35 / 25 / 0$

Hence, $5.50 \times 4.25 = 23.3750 = 23.375$

This clearly shows that fractions can be easily solved by converting them into decimal fractions.

Example: Multiply $3\ 4/7$ by $12\ 5/8$.
Solution: Since we can't convert the fractional part 4/7 into a terminating decimal fraction, it is advisable to convert both the mixed fractions into improper fractions at this stage.

$$3\ 4/7 \times 12\ 5/8 = 25/7 \times 101/8 = \frac{25 \times 101}{7 \times 8} = \frac{2525}{56}$$

Division of Fraction

Dividing fractions is as easy as pie! Change the mixed fraction into improper fraction first. Flip the numerator and denominator of the second fraction and multiply the two fractions. The simple rule is known as KCF (Keep Change Flip).

If you have

$$\frac{a}{b} \div \frac{c}{d}$$

Keep Change Flip

$$\frac{a}{b} \times \frac{d}{c}$$

Example: Find the value of $1/4 \div 3/8$
Solution: Apply the KCF method and the answer is,
$1/4 \times 8/3 = 8/12 = 2/3$

Here's another—

Example: Find the value of $3\ 1/2 \div 4\ 1/8$
Solution: Here we shall apply the circle method to change the mixed fraction into an improper fraction.

What is the circle method? A method to change a mixed fraction into improper fraction.

$$4\frac{1}{3} = \frac{4 \times 3 + 1}{3} = \frac{13}{3}$$

Hence, $3\ 1/2 \div 4\ 1/8 = 7/2 \div 33/8$
$= 7/2 \times 8/33 = 28/33$

Case 1: To divide by mixed fraction $2\ 1/3$, $3\ 1/4$, $4\ 1/5$, etc.

If the mixed fraction has $1/3, 1/4, 1/5, 1/6$ as fractions, then multiply the whole parts with 3, 4, 5, 6, etc., and do the divisions as usual.

Example: $12 \div 2\ 1/3$
Solution: $12 \times 3 = 36$
And $3 \times 2\ 1/3 = 7$
Hence, $12 \div 2\ 1/3 = 36/7$

Example: $24 \div 5\frac{1}{6}$
Solution: $24 \times 6 = 144$
And $6 \times 5\frac{1}{6} = 31$
Hence, $24 \div 5\frac{1}{6} = 144/31$

Case 2: To divide a number by $7\frac{1}{2}$

Multiply both the dividends and divisor with 4 and divide the result.

Example: $12 \div 7\frac{1}{2}$
Solution: $12 \times 4 = 48$
And $4 \times 7\frac{1}{2} = 30$
Hence, $12 \div 7\frac{1}{2} = {}^{48}/_{30} = {}^{24}/_{15}$

Example: $128 \div 7\frac{1}{2}$
Solution: $128 \times 4 = 512$
And $4 \times 7\frac{1}{2} = 30$
Hence, $128 \div 7\frac{1}{2} = {}^{512}/_{30} = {}^{256}/_{15}$

Case 3: To divide a number by $12\frac{1}{2}$

Multiply both the dividends and divisor with 8 and divide the result.

Example: $12 \div 12\frac{1}{2}$
Solution: $12 \times 8 = 96$
And $8 \times 12\frac{1}{2} = 100$
Hence, $12 \div 12\frac{1}{2} = {}^{96}/_{100} = {}^{24}/_{25}$

Example: $125 \div 12\frac{1}{2}$
Solution: $125 \times 8 = 1000$
And $8 \times 12\frac{1}{2} = 100$
Hence, $125 \div 12\frac{1}{2} = {}^{1000}/_{100} = 10$

Questions on simple fractions involving addition, subtraction, multiplication and division come in most of the competitive examinations and the golden rules discussed above will certainly help you solve such questions on time. The best thing you can do is to master the rules with practice.

Practice Problems

Add:
a) $1/7 + 1/8 + 5/14$
b) $3/5 + 2/8$
c) $2\,1/4 + 3\,1/7$
d) $4\,1/6 + 5\,1/9$

Subtract:
a) $11/8 - 3/7$
b) $2/11 - 7/13$
c) $2\,4/7 - 1\,1/8$
d) $1\,5/6 - 1\,1/9$

Multiply:
a) $1\,2/7 \times 3\,1/4 \times 4\,1/7$
b) $2/13 \times 13/9 \times 34/8$
c) $5/17 \times 1\,2/7 \times 3\,4/7$
d) $4\,5/9 \times 2\,3/4$

V
GOLDEN RULES TO YOUR 'SPEEDY' RESCUE

Casting Out Elevens vs Casting Out Nines

Casting Out Nines and Casting Out Elevens are ways to check the validity of a calculation. Most of the fundamental arithmetical operations can be cross-checked with these two methods, whether it is an addition, subtraction, multiplication or division. They can also be extended to check a square, square root, cube and cube root. The basic difference between these two methods can be seen in the examples. It is for you to decide which method is the best.

Casting Out Nines

Casting Out Nines or the Chinese Remainder Theorem is a method where we find the digit sum of the given number by removing the number 9 or the digit sum 9. The final sum of the number should not exceed 8.

Example: Find the digit sum of 4379348568219.
Solution: Add all digits. Group those numbers that have the digit sum of 9.

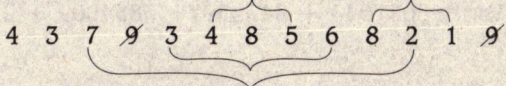

Here, $3 + 6 = 8 + 1 = 4 + 5 = 7 + 2 = 9$ has been cancelled out. Moreover, there are two 9s that will also be left out. The digit sum of the left-out digits is—

$$4 + 3 + 8 = 15 > 9$$
$$1 + 5 = 6$$

Example: Find the digit sum of 4954653.

Solution: Digit sum of 4954653 =

$$4 + \cancel{9} + \underbrace{5 + 4} + \underbrace{6 + 5} + 3 = 0$$

When applied to addition

Check the following addition: $123 + 456 + 789 + 912 = 2280$

Solution: Using the Casting Out Nines' method, first find the digit sum of each number and write them against that number.

Digit sum

```
1 2 3        6
4 5 6        6  | 6+6 = 12 > 9
7 8 9        6  | 1+2 = 3
9 1 2        3
─────       ───
2280         3
```

The first digit sum is written against the number.

Digit sum of the final answer $2280 = 2 + 2 + 8 + 0 = 12$; $1 + 2 = 3$

Since the digit sum of the number written against the number is also coming out to be 3 which is equal to the digit sum of the final result, the answer is correct.

Example: Verify $5087643 + 8432397 + 3854009 + 2197565 = 19571614$

Verification:

	Digit Sum of Number
5087643	6
8432397	0
3854009	2
2197565	8
19571614	?

LHS* = Digit sum of 19571614 = 1 + 9 + 5 + 7 + 1 + 6 + 1 + 4 = 7

RHS = Sum of the digit sum of number = 6 + 0 + 2 + 8 = 7

Result verified.

Example: 10045 + 34567 + 88888 + 234 = 145734

Verification:

	Digit Sum of Number
10045	1
34567	7
88888	4
234	0
145734	?

LHS = Digit sum of 145734 = 1 + 4 + 5 + 7 + 3 + 4 = 6

RHS = Sum of the digit sum of number = 1 + 7 + 4 + 0 = 3

Since, LHS ≠ RHS, result is incorrect.

When applied to subtraction

Let's check two examples to understand how, when applied to subtraction problems, the Casting Out Nines works.

---

*LHS is 'left hand side' and RHS, 'right hand side'.

Example: Subtract 2458 from 4612.
Solution:

$$
\begin{array}{r} & \text{Digit Sum} \\ 4612 & 4 \to 4-1=3 \\ -2458 & -1 \\ \hline 2154 & =3 \end{array}
$$

As you can see, the difference between the digit sum of 4612 and 2458 comes out to be the digit sum of 2154. It makes the subtraction correct and the process of checking using Casting Out Nines makes the confirmation of the calculation easy.

Example: Verify $3456928734 - 1958762087 = 1498166647$.
Solution:

	Digit Sum of Number
3456928734	6
−1958762087	8
1498166647	?

LHS = Digit sum of $1498166647 = 1 + 4 + 9 + 8 + 1 + 6 + 6 + 6 + 4 + 7 = 7$

RHS = $6 - 8 = -2$

Since, the result is negative, add 9 to it to make it positive. Hence, RHS = $-2 + 9 = 7$

LHS = RHS

Hence, result verified.

Moreover, you can write $6 - 8$ as $60 - 8$ or $15 - 8$ or $24 - 8$ or $33 - 8$, as a digit sum of 60, 15, 33, 24 are same.

When applied to multiplication

In multiplication too, the same process can be used. We all know:
Multiplicand × Multiplier = Result

Now, let's check out a few examples.

Example: Verify 2467 × 8797 = 21702199.
Check: Digit sum of 2467 = 1

Digit sum of 8797 = 4

LHS = Digit sum of 2467 × Digit sum of 8797 = 1 × 4 = 4

RHS = Digit sum of 21702199 = 4

Since, LHS = RHS, result is verified.

Example: Verify 85942 × 3054 = 262460868.
Solution: Digit sum of Multiplicand = 8 + 5 + 9 + 4 + 2 = 1

Digit sum of Multipliers = 3 + 0 + 5 + 4 = 3

Digit sum of Results = 2 + 6 + 2 + 4 + 6 + 0 + 8 + 6 + 8 = 6

LHS = Digit sum of Multiplicand × Digit sum of Multiplier = 1 × 3 = 3

RHS = Digit sum of Result = 6

LHS ≠ RHS

Hence, result is incorrect.

Why casting out elevens is better than casting out nines

Casting Out Elevens has an advantage over Casting Out Nines because the latter method fails to differentiate between 125, 251 and 215 as the digit sum of 125, 251 and 215 is 8. Let's take an example to understand the flaws of the Casting Out Nines method.

Example: Add 452, 621, 970 and 167.
Solution: Let's see the working of two students and try to check which one is correct.

452	452
621	621
970	970
+167	+167
2210	2120

Here, digit sum of answer 2210 and 2120 is 5. In such a case, the Casting Out Nines method fails to judge the correct answer. Let's take another example.

Example: 14 × 12 = ?
Solution:

I	II	III
14	14	14
×12	×12	×12
168	681	861

In the above multiplication problems, the final answers in three cases are different though the digit sum of 168, 681 and 861 are the same.

In such a scenario, Casting Out Elevens can come in handy as it understands the difference between 168, 681 and 861. Now let's see how Casting Out Eleven works.

Casting Out Elevens

Before I tell you the working of Casting Out Elevens, let me tell you the divisibility rule of 11 first.

a) Find the sum of digits at even places and odd places separately.
b) Find the difference between these sums.
c) If the difference is either 0 or 11, then the number is divisible by 11.

Example: Are 542, 781 and 192 divisible by 11?
Solution: Sum of digits at even places = 9 + 1 + 7 + 4 = 21
Sum of digits at odd places = 2 + 1 + 8 + 2 + 5 = 18
Difference = 21 − 18 = 3
Hence, 542, 781 and 192 are not divisible by 11.

Example: Is 14641 divisible by 11?
Solution: Sum of digits at odd places = 1 + 6 + 1 = 8
Sum of digits at even places = 4 + 4 = 8
Difference = 8 – 8 = 0
Hence, 14641 is divisible by 11.

Let's do something different.

- Write the number in reverse order.
- Put plus (+) or minus (–) alternately between the digits.
- If the result obtained is divisible by 11 then the number is divisible by 11.

Example: Is 14641 divisible by 11?
Solution: 1 – 4 + 6 – 4 + 1 = 0
Hence, it is divisible by 11.

Example: Is 542781192 divisible by 11?
Solution: 2 – 9 + 1 – 1 + 8 – 7 + 2 – 4 + 5
If the result is negative, add 11. So,
–3 + 11 = 8 is not divisible by 11.

Now, let's check some mathematical operations when Casting Out Elevens is applied to them.

When applied to addition

Example: Is 245 + 289 = 534?
Solution
245: 5 – 4 + 2 = 3
289: 9 – 8 + 2 = 3
534: 4 – 3 + 5 = 6
Here, 3 + 3 = 6
Hence it is correct.

Let's change the final result.

Example: Is $245 + 289 = 543$?
Solution

245: $5 - 4 + 2 = 3$
289: $9 - 8 + 2 = 3$
543: $3 - 4 + 5 = 4$
Here, $3 + 3 \neq 4$
So, the sum total is wrong.

The basic difference between the methods of Casting Out Nines and Casting Out Elevens makes a big difference in calculating. In Casting Out Nines, the order of digits in a number doesn't matter and gives you the same result, but the Casting Out Elevens method differentiates the order of digits in a number. In that way, the latter method is more advantageous than the Casting Out Nines method. Let's look at the following explanation.

Casting Out Nines	Casting Out Elevens
Digit sum of $634 = 6 + 3 + 4 = 4$	$634 : 4 - 3 + 6 = 7$
Digit sum of $463 = 4 + 6 + 3 = 4$	$463 : 3 - 6 + 4 = 1$
Digit sum of $643 = 6 + 4 + 3 = 4$	$643 : 3 - 4 + 6 = 5$

When applied to subtraction

Let's check out some subtraction problems using Casting Out Elevens. Here are two examples.

Example: $4127 - 2376 = 1751$
Solution: First find the Casting Out Elevens values of each term.

4127: $7 - 2 + 1 - 4 = 2$
2376: $6 - 7 + 3 - 2 = 0$
Hence, $4127 - 2376 = 2 - 0 = 2$
Casting Out Elevens values for $1751 = 1 - 5 + 7 - 1 = 2$
This shows that the answer is correct.

Example: $6541 - 4879 = 1626$
Solution: First find the Casting Out Elevens values of each term.

6541: $1 - 4 + 5 - 6 = -4 + 11 = 7$
4879: $9 - 7 + 8 - 4 = 6$
Hence, $6541 - 4879 = 7 - 6 = 1$
Casting Out Elevens values for $1626 = 6 - 2 + 6 - 1 = 9$
Since, $1 \neq 9$
Hence, answer is incorrect.

When applied to multiplication

Example: Is $243 \times 257 = 62451$?
Solution

243: $3 - 4 + 2 = 1$
257: $7 - 5 + 2 = 4$
62451: $1 - 5 + 4 - 2 + 6 = 4$

Since, $1 \times 4 = 4$
Hence, the multiplication is correct.

Example: Is $4672 \times 3469 = 16207168$?
Solution

4672: $2 - 7 + 6 - 4 = -3 + 11 = 8$
3469: $9 - 6 + 4 - 3 = 4$

Now, $4672 \times 3469 = 8 \times 4 = 32$

32: $2 - 3 = -1 + 11 = 10$

Result = 16207168: $8 - 6 + 1 - 7 + 0 - 2 + 6 - 1 = -1 + 11 = 10$

Since both sides are equal, the result is verified.

When applied to division

To check division, we use the following procedure:

1. Subtract the remainder, if any, from the actual dividend. Find the elevens remainder of the dividend or reduced dividend.
2. Find the elevens remainder of the divisor.

3. Find the elevens remainder of the quotient.
4. Use the formula: Dividend = Divisor × quotient + Remainder

Let's check out the following examples.

Example: Check 4274 ÷ 24; Quotient = 178 and Remainder = 2
Solution: Casting Out Elevens for Dividend
4274: 4 − 7 + 2 − 4 = − 5 + 11 = 6
Casting Out Elevens for Divisor
24: 4 − 2 = 2
Casting Out Elevens for Quotient
178: 8 − 7 + 1 = 2
Casting Out Elevens for Remainder
2: 2

Apply the formula: Dividend = Divisor × Quotient + Remainder
RHS = Divisor × Quotient + Remainder
= 2 × 2 + 2
= 6
LHS = 6
Hence, result is verified.

Example: Check 73521 ÷ 27; Quotient = 2723; Remainder = 0
Solution: Let's find the Casting Out Elevens remainder of each.
Casting Out Elevens of Dividend 73521: 1 − 2 + 5 − 3 + 7 = 8
Casting Out Elevens for Quotient 2723: 3 − 2 + 7 − 2 = 6
Casting Out Elevens for Divisor 27: 7 − 2 = 5

Apply the formula: Dividend = Divisor × Quotient + Remainder
RHS = Divisor × Quotient + Remainder
= 5 × 6 + 0
= 30
Casting Out Elevens of RHS 30 = 3 − 0 = 3
Casting Out Elevens of Dividend 27 = 7 − 2 = 5
Since LHS ≠ RHS, result is incorrect.

Hope you have understood the difference between Casting Out Nines and Casting Out Elevens? Try out both and see for yourself!

Practice Problems

Verify the following results by using the Casting Out Elevens as well as Casting Out Nines methods:

a) 112065 + 360085 + 289872 + 156345 = 918367
b) 4998 + 6789 + 5715 + 4837 + 8976 = 31315
c) 7534 + 2459 + 1932 + 6547 = 16472
d) 658723 + 154639 − 369847 + 367 = 443882
e) 3746735 − 2837546 = 909189
f) 876542 − 32548 − 698547 = 145447
g) 588 × 512 = 301056
h) 842 × 858 = 722536
i) 13579 ÷ 975, Q = 13, R = 904
j) 7238761 ÷ 524, Q = 13184, R = 225
k) 87265 × 32117 = 2802690005

Rule of 72

In simple interest, you must have solved questions like 'In how many years will a sum get doubled if the effective rate of interest is 12% per annum?'

Rule of 72 helps the investor estimate the time period/ rate of interest at which the given amount is doubled. Earlier, banks and post offices used to announce doubling the money after certain periods and if you were an investor, you could estimate the time period with the help of rule of 72. Let's see the example.

If you have kept some amount in a bank that gives you the interest of 10% compounded annually, then it will take 72/10 = 7.2 years to get the amount doubled.

This is not the actual method but is useful for mental calculations.

The actual time to get a sum doubled at the rate of 10% is 7.3 years but the estimation of 7.2 in your head without proper calculation, is not bad at all.

Similarly, if you want your money to get doubled in 5 years, then the rate of interest should be = 72/5 = 14.4 years or approximately 15 years.

Here is a table that will help you to understand the actual picture.

Rate of Interest	Time with Rule of 72	Actual Time	Difference of Time
2%	36.0	35	1.0
3%	24.0	23.45	0.6
5%	14.4	14.21	0.2
7%	10.3	10.24	0.0
9%	8.0	8.04	0.0
12%	6.0	6.12	0.1
25%	2.9	3.11	0.2
50%	1.4	1.71	0.3

As you can see, as the rate of interest grows, the difference between the actual time and the estimated time comes out to be minimal.

Now the question is why 72 is taken as base for the estimation.

72 has many factors—2, 3, 4, 6, 8, 9, 12, etc. This makes it convenient to use 72 as an indicator. Though rule of 69 or 69.3 is far better than 72, calculating divisions in one's mind with a denominator of 69.3 is not easy.

There is yet another rule that gives the correct time to double the money; for the rates from 0% to 20%. This rule is called E-M rule or Eckart–McHale second order rule. Also called the rule of 69.3, it is considered to be better than the rule of 72 since it gives the accurate answer if the interest rate is between 0% and 20%. Here is the formula to calculate the time with E-M rule.

$$\text{Time} = \frac{69.3}{r} \times \frac{200}{200-r}$$

Example: In how many years will a sum be doubled at the rate of 9%?

Solution: By Rule of 72, Time = 72/9 = 8 years

By E-M rule,

Time = $\dfrac{69.3 \times 200}{9 \times (200 - 9)}$

= $\dfrac{69.3 \times 200}{9 \times 191}$

= 8.06 years

You can see the slight difference between the two calculations but the difference is minimal in view of the calculation done in one's mind.

Example: In how many years will a sum be doubled at the rate of 12.5%?

Solution: By Rule of 72, Time = 72/12.5 = 5.76 years

By E-M rule,

Time = $\dfrac{69.3 \times 200}{12.5 \times (200 - 12.5)}$

= $\dfrac{69.3 \times 200}{12.5 \times 187.5}$

= 5.91 years

Example: In how many years will a sum be doubled at the rate of 6%?

Solution: By Rule of 72, Time = 72/6 = 12 years

By E-M rule,

Time = $\dfrac{69.3 \times 200}{6 \times (200 - 6)}$

= $\dfrac{69.3 \times 200}{6 \times 194}$

= 11.91 years

Practice Problems

Applying E-M Rule and Rule of 72, find the time taken to double the money invested in a bank in the following instances:

a) p = 1200   r = 6%

b) p = 2628   r = 4.5%
c) p = 2500   r = 12.5%
d) p = 4000   r = 16.2%
e) p = 11000   r = 18.7%

Finding Prime Numbers

In competitive examinations, you are asked to find the average of the first 10 prime numbers or to find the prime numbers between 1 to 100, etc. Many a times, you are asked to test whether the given number is a prime number or not? To answer such questions, you need to first know what a prime number is.

What is a Prime Number?

A prime number is an integer p which is not 0 or ± 1 and is divisible by no integer except ±1 and itself.

2 is the only even prime number and the rest of the prime numbers known so far are odd. An even number is one if it is divisible by 2.

Example: 2, 4, 6, 8, etc.

On the other hand, if a number is not divisible by 2 or can be written in the form of 2m + 1, then the number is an odd number.

Example: 1, 3, 5, 7, 9, etc.

There is no perfect technique that can immediately help us find the prime number between two numbers. Eratosthenes (276–194 BCE), a great Greek mathematician, suggested a method called Sieve of Eratosthenes. This golden rule, though simple, is time-consuming. It states: 'First write down the number from 2 to N. Remove all the multiples of 2, 3 and continue the process until all the multiple of primes not greater than √N has been removed.'

Test of Prime Number

Eratosthenes described a method to find the prime number between two given numbers. In competitive examinations, you will be asked to test whether a given number is a prime number or not. Moreover, many a times questions like how many primes are there between 1 to 50 or 50 to 100 or 1 to 100 do come in such examinations and therefore I would advise you to remember the primes below 100 so that you can quickly answer the question.

Rule: If you are asked to find the primes between 1 to 100 then first write 1 to 100. The square root of 100 is 10, so circle all primes below 10; they are 2, 3, 5 and 7. Cancel out all the multiples of 2, 3, 5 and 7. The remaining numbers are primes.

Remember, 1 is not a prime number.

Question: Why is 1 not a prime number?
Answer: As per the Fundamental Theorem of Arithmetic or Unique Factorization Theorem, 'Every positive whole number can be written as a unique product of primes.'

So, 1 can be written as
$1 = 1 \times 1$
$\quad = 1 \times 1 \times 1$
$\quad = 1 \times 1 \times 1 \times 1 \times 1$

This contradicts the basic principle of the Factorization Theorem. Hence, 1 is a special number which is neither a prime nor a composite.

Now, let's return to our main question of finding primes between two given numbers. First try to find the primes between 1 and 100. Since 1 is not a prime number, first cut it out. Now begin with the first prime number, 2. Encircle it and cut out all the multiples of 2. Repeat the operation with 3, 5 and 7 as

$\sqrt{100} = 10$ so we need to circle all primes below 10. The numbers that are left, including those encircled, are Prime Numbers.

1	2	3	4	5	6	7	8	9	10
11	12	13	14	15	16	17	18	19	20
21	22	23	24	25	26	27	28	29	30
31	32	33	34	35	36	37	38	39	40
41	42	43	44	45	46	47	48	49	50
51	52	53	54	55	56	57	58	59	60
61	62	63	64	65	66	67	68	69	70
71	72	73	74	75	76	77	78	79	80
81	82	83	84	85	86	87	88	89	90
91	92	93	94	95	96	97	98	99	100

In examinations, many a times you will be asked to find the mean, median, etc., of primes between two given numbers. You can accurately answer such questions provided you know the prime numbers.

So, there are 25 prime numbers less than 100, 2 being the smallest and 97, the highest prime number below 100. Prime numbers below 100 are 2, 3, 5, 7, 11, 13, 17, 19, 23, 29, 31, 37, 41, 43, 47, 53, 59, 61, 67, 71, 73, 79, 83, 89 and 97.

Suppose you are asked that for a given number **p**, which you are going to test for prime, find a number **n** such that $n^2 \geq p$. Now divide **p** with all primes less than **p** and if **p** is not divisible by either of them, then **p** is definitely a prime number, else it is a composite number.

Example: Is 437 a prime number?
Solution: Here $(21)^2 > 437$ and prime numbers below 21 are 2, 3, 5, 7, 11, 13, 17, 19 and 437 is divisible by 19. Hence it is not a prime number.

Example: Is 517 a prime number?
Solution: Here $(23)^2 > 517$ and prime numbers below 23 are 2, 3, 5, 7, 11, 13, 17 and 19. The given number 517 is divisible by 11. Hence it is not a prime number.

Example: Is 811 a prime number?
Solution: Here $(30)^2 > 811$ and prime numbers below 30 are 2, 3, 5, 7, 11, 13, 17, 19, 23 and 29. Interestingly, 811 is not divisible with any of the prime numbers given here. Hence 811 is a prime number.

Practice Problems

1. Find the mean of the first five prime numbers.
2. Find the median of the first ten prime numbers.
3. Which of the following is a prime number?
   a) 914   b) 8751
   c) 1093   d) 10987
   e) 47   f) 679

Finding the Unit Digit

Find the unit digit of $(2467)^{153} \times (341)^{72}$.

If one intends to solve this question with a simple method then I doubt if even a calculator will give you the exact answer. Let's therefore try out something new and speedy. To understand the concept of unit digit, you need to understand the concept of a cycle and draw a pattern.

Case 1: Let's first begin with 2.

$2^1 = 2$
$2^2 = 4$
$2^3 = 8$
$2^4 = 16$
$2^5 = 32$

The above pattern shows that the unit digit 2 repeats on the fifth power, as the unit digit of 2^5 is same as the unit digit of 2^1. It means that the cycle of 2 is 4 as after every fourth multiplication, the unit digit will be 2. Hence the cycle of 2 has 4 different numbers: 2, 4, 8 and 6.

Let's check out the cycle of 3.

$3^1 = 3$
$3^2 = 9$
$3^3 = 27$
$3^4 = 81$
$3^5 = 243$

Hence you can conclude that the cycle of 3 has 4 different numbers at the unit place and they are 3, 9, 7 and 1. This tells you that the cycle of 3 is similar to 2.

Now check out the cycle of 7.

$7^1 = 7$
$7^2 = 49$
$7^3 = 343$
$7^4 = 2401$
$7^5 = 16807$

This means number 7 has a cycle of 4, as both 7^5 and 7^1 end with 7 each. Similarly, the number 8 has the cycle of 4.

Note: 2, 3, 7 and 8 have a unit digit cycle of four.

Case 2: Unit digit of 4 and 9.

Let's first check the cycle of 4.

$4^1 = 4$
$4^2 = 16$
$4^3 = 64$
$4^4 = 256$
$4^5 = 1024$

You can observe that the power cycle of 4 contains only 2 numbers—4 and 6. Hence the cycle of 4 is 2. Moreover,
$(4)^{odd} = 4$ and $(4)^{even} = 6$

Likewise, the cycle of 9 is also 2. Check out the following:

$9^1 = 9$
$9^2 = 81$
$9^3 = 729$

The power cycle here too contains only two numbers: − 9 and 1.
Here, you can draw two conclusions:
$(9)^{even} = 1$ $\qquad\qquad$ $(9)^{odd} = 9$

Finding the Unit Digit

So far we have seen how a number that ends with 2, 3, 4, 7, 8 and 9 behaves and now we are left with 0, 1, 5 and 6. Let's see how to find the unit digit of a number that ends with 0, 1, 5 and 6.

Let's begin with the different power of 0, 1, 5 and 6.

$0^2 = 0$, $0^4 = 0$, $0^9 = 0$
$1^2 = 1$, $1^6 = 1$, $1^9 = 1$ ...
$5^1 = 5$, $5^2 = 25$, $5^5 = 3125$ ...
$6^1 = 6$, $6^2 = 36$, $6^3 = 216$ ...

The conclusion that you can draw from the above is that they all have the same unit digit as the number itself with any power raised to these numbers. Let's summarize the above discussion in a table.

Number	Cycle
1	1
2	4
3	4
4	2
5	1
6	1
7	4
8	4
9	2

Try to remember this table for speedy calculation of questions that ask you to find the unit digit of a number. Now, let's look at the following examples.

Example: Find the unit digit of the following:
 a) 269^{145789}
 b) 19^{148}

Solution:
 a) $(9)^{odd} = 9$
 b) $(9)^{even} = 1$

Example: Find the unit digit of $(2018)^{2018}$
Solution: $(2018)^{2018} = (2018)^{4 \times 504 + 2}$

Since the cycle of numbers ending with 8 is 4, power 2018 is broken into the power of 4.

$2018 = 4 \times 504 + 2$

Hence, simply find $(8)^2 = 64$

Therefore, the unit digit of $(2018)^{2018} = 4$

Example: Find the unit digit of $(5615)^{2567298}$
Solution: As described above, $(5)^n$ ends with 5.

Hence, unit digit of $(5615)^{2567298} = 5$

Example: Find the unit digit of $(154)^{258741369}$
Solution: We know, $(4)^{odd} = 4$

Hence, unit digit of $(154)^{258741369} = 4$

Let's try another method which is nothing but the simplification of the above method. First, the rules: Divide the exponent by 4 and find the remainder.

Unit digit of number = (Unit digit)^{remainder}

Example: Find the unit digit of $(1453)^{71}$
Solution: Divide exponent by 4.

$71 \div 4$; Q = 17 and R = 3

Hence, unit digit of $(1453)^{71}$ = Unit digit of $(3)^3 = 7$

Example: Find the unit digit of $(2016)^{2015} - (2015)^{2016}$
Solution: $2015 \div 4$; Q = 503; and R = 3

Unit digit of $(2016)^{2015}$ = Unit digit of $(6)^3 = 6$

And $2016 \div 4$; Q = 504 and R = 0

Unit digit of $(2015)^{2016}$ = Unit digit of $(5)^0 = 1$

Hence, the unit digit of $(2016)^{2015} - (2015)^{2016} = 6 - 1 = 5$

Example: Find the unit digit of $(2)^{51}$
Solution: Divide exponent by 4.

$51 \div 4$; Q = 12 and R = 3

Hence, unit digit of $(2)^{51}$ = Unit digit of $(2)^3 = 8$

Let's summarize the chapter.

- If unit digit of a number whose exponent value is to be finalized is 2, 4, 6 and 8, then unit digit of number will be 6.
- If unit digit of a number whose exponent value is to be finalized is 3, 7 and 9, then unit digit of number will be 1.
- If unit digit of a number whose exponent value is to be finalized is 1, 5 and 6, then unit digit of number will be 1, 5 and 6 respectively.

Practice Problems

Find the unit digit of the following:

a) $(114)^5 \times (467)^{201}$
b) $(9865)^{754} \times (7204)^{149}$
c) $(12986754)^{2019}$
d) $(47772)^{9201} \times (65521)^{9801}$
e) $(217)^9 \times (128)^{11} \times (76622)^{24}$

The Magic of the Pascal Triangle

First discovered in India and named as 'Meru Prastha', it was, however, Blaise Pascal, a French mathematician, who explored the Pascal Triangle and made it popular. The latter used this triangle to solve the binomial expansion of **n** degree without using the concept of permutation and combination, as the coefficient of each term of expansion of power **n** could be seen in the triangle itself.

The Pascal Triangle is a tool used to find the expansion of an algebraic identity or expression. Moreover, it is also an effective tool to understand the different mathematical properties.

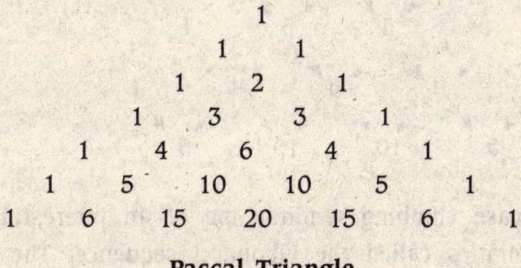

Pascal Triangle

How It Works

Have you put candles on a staircase during Diwali? If yes, then you can make this triangle on your own. First put 1 diagonally in two directions as shown below.

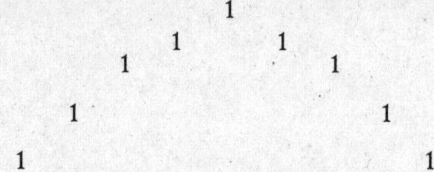

Now add two numbers from consecutive rows and write the results in rows on the next level.

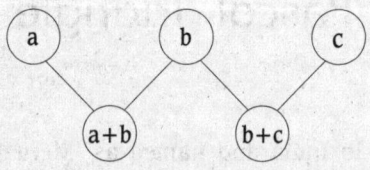

This is the way to complete the triangle. For your convenience, five rows have been completed here. You can make a larger triangle if you wish.

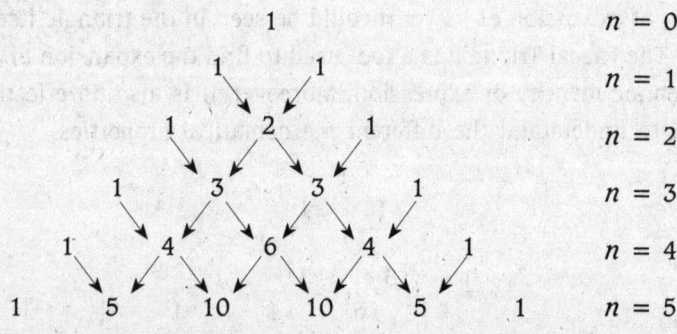

The staircase climbing reminds me of an interesting pattern in mathematics called the Fibonacci sequence. The Fibonacci sequence of numbers is seen everywhere in nature. 1, 1, 2, 3, 5, 8, 13, 21, 34, etc., is a Fibonacci sequence where every number is the sum of the two preceding numbers.

$2 = 1 + 1$
$3 = 1 + 2$
$5 = 2 + 3$

Now, come to the pattern of climbing the staircase. If you climb the staircase by jumping 1 step or 2 steps at a time then you can notice the following pattern.

n	Ways to Climb Stairs	a_n
1	1	1
2	11, 2	2
3	111, 12, 21	3
4	1111, 112, 121, 211, 22	5
5	11111, 1112, 1121, 1211, 2111, 122, 212, 221	8

The ways to climb stairs is a Partition number pattern that was discovered by Srinivasa Ramanujan.

$P(5) = 1 + 1 + 1 + 1 + 1 = 1 + 1 + 1 + 2 = 1 + 1 + 2 + 1 = 1 + 2 + 1 + 1 = 2 + 1 + 1 + 1 = 1 + 2 + 2 = 2 + 1 + 2 = 2 + 2 + 1$

The total number of ways is written in the third column that is a Fibonacci sequence.

Uses and Features of a Pascal Triangle

The Pascal Triangle saves students from the burden of mugging up formulae. First look at the power (n) of expansion,

- Write the coefficient from the row equal to the power of expansion. If n = 3, then move the third row from the top (leaving the top 1).
- Put the variable with each of the coefficients.
- Start from the left and give the maximum power (n) to the first variable and 0 to the second variable. Keep decreasing the power of the first variable by 1 and increase the power of the second variable by 1 in each successive term.
- Place a plus sign in between each term. If the expansion has a negative sign between the variables, place plus, minus signs alternately between each successive term.

The Magic of the Pascal Triangle

Example: Write the expansion of $(m + n)^2$

Solution: Power (n) = 2

Expansion of second row = 1 2 1

Put variable m and n with each of the coefficients.

                1 mn   2 mn   1 mn

Give maximum power to the first variable and 0 to the second. Decrease the power of the first variable and increase the power of the second variable by 1.

                $1\ m^2n^0$ $2\ m^1n^1$ $1\ m^0n^2$

Since there is a plus sign between variables in the expansion of $(m + n)^2$, we need to put a plus sign in between each successive term.

$(m + n)^2 = 1\ m^2n^0 + 2\ m^1n^1 + 1\ m^0n^2$
$= m^2 + 2\ mn + n^2$

Example: Write the expansion of $(m + n)^3$

Solution: Power (n) = 3

Expansion of third row = 1 3 3 1

Put variable m and n with each of the coefficients.

            1 mn   3 mn   3 mn   1 mn

Give maximum power to the first variable and 0 to the second. Decrease the power of the first variable and increase the power of the second variable by 1.

      $1\ m^3n^0$ $3\ m^2n^1$ $3\ m^1n^2$ $1\ m^0n^3$

Since there is a plus sign between variables in the expansion of $(m + n)^3$, we need to put a plus sign in between each successive term.

$(m + n)^3 = 1\ m^3n^0 + 3\ m^2n^1 + 3\ m^1n^2 + 1\ m^0n^3$
$= m^3 + 3\ m^2n^1 + 3\ m^1n^2 + n^3$

Example: Expand $(m - n)^4$

Solution: Look at the Pascal Triangle at the top. Note down the coefficients from the fourth row and they are—1, 4, 6, 4 and 1. As explained in the two examples above, the exponents will start at m^4n^0 and move to m^3n^1, m^2n^2, etc. Moreover, there is a negative sign between the variables; hence alternate signs of plus and minus will appear.

$$(m - n)^4 = 1\ m^4n^0 - 4\ m^3n^1 + 6\ m^2n^2 - 4\ m^1n^3 + 1m^0n^4$$
$$= m^4 - 4\ m^3n + 6\ m^2n^2 - 4\ mn^3 + n^4$$

Isn't it easy to expand?

Let's check out some more expansions.

Example: Find the expansion of $(2 + \sqrt{5})^5$

Solution: From the triangle, we can first write the coefficient on the fifth row—1, 5, 10, 10, 5, and 1.

Now, as explained, write the answer in a single line.

$$(2 + \sqrt{5})^5 = 1\ (2)^5 + 5\ (2)^4\ (\sqrt{5})^1 + 10\ (2)^3\ (\sqrt{5})^2 + 10\ (2)^2\ (\sqrt{5})^3 + 5\ (2)^1\ (\sqrt{5})^4 + 1\ (2)^0\ (\sqrt{5})^5$$
$$= 32 + 80\sqrt{5} + 400 + 200\sqrt{5} + 250 + 25\sqrt{5}$$
$$= 682 + 305\sqrt{5}$$

This shows that if you have enough practice, you need not remember the algebraic identity. In many competitive examinations, you are asked to find the nth term from a given expansion. You can do it without expanding the whole.

For the given expansion—$(a + b)^n$

$$r\text{th term} = \frac{n!}{(n - r + 1)!\ (r - 1)!} a^{n - r + 1} b^{r - 1}$$

If the binomial has a positive sign, all terms will be positive. In case there is a negative sign, the term will be negative if r is even, whereas the term will be positive if r is odd.

Example: Find the 12th term of $(x - y)^{14}$

Solution: Here $a = x$, $b = y$, $n = 14$ and $r = 12$

Since $r = 12$ is even, the 12th term will be negative.

Hence, 12th term =

$$= -\frac{14!}{(14 - 12 + 1)!\ (12 - 1)!} x^{14-12+1} y^{12-1}$$
$$= -\frac{14!}{3!\,11!} x^3 y^{11}$$
$$= -364 x^3 y^{11}$$

I do hope you have understood the importance of the Pascal Triangle in doing calculations.

But could you tell in what other ways the Pascal Triangle can help you?

The Triangle itself is full of different kinds of numbers. Let's check them out.

a) Write the coefficient as in the Pascal Triangle in the following manner.

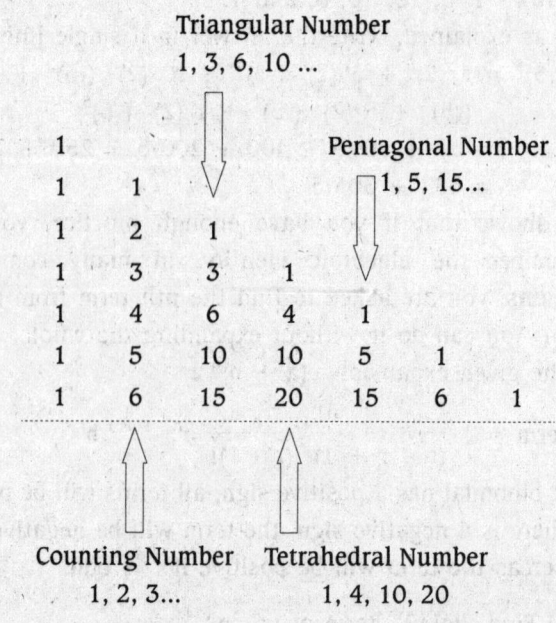

b) If you add the coefficient of digits in each row and write them against that row, you will find a beautiful pattern.

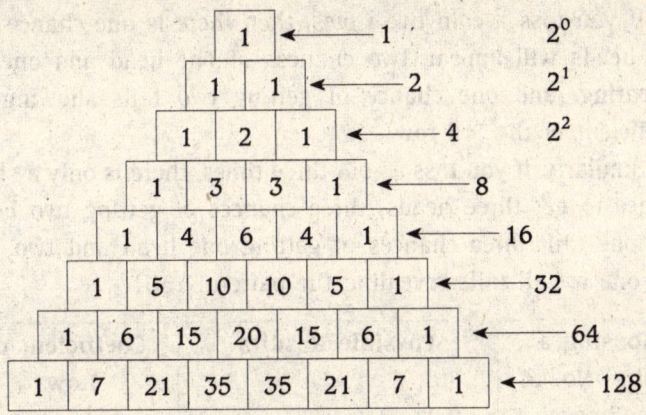

The sum of the 3rd row = 2^2 = 4
The sum of the 4th row = 2^3 = 8
The sum of the 5th row = 2^4 = 16
Hence, the sum of the 8th row is 2^7 = 128

Here, 1, 2, 4, 8, 16, 32, 64 and 128 form a geometric progression series and the sum of nth term of a GP is given by:

$$S_n = \frac{a(r^n - 1)}{r - 1}$$

Let's see another feature of the Pascal Triangle.
In the 2nd row we have the coefficient = 1 1
In the 3rd row the coefficient = 121 = $(11)^2$
In the 4th row the coefficient = 1331 = $(11)^3$
In the 5th row the coefficient = 14641 = $(11)^4$
In the 6th row the coefficient = 15101051 = 1 / 5 + 1 / 0
+ 1 / 0 / 5 / 1
= 161051

The most important use of Pascal triangle is that it can show you how many times heads and tails can appear when tossing a coin. If you toss a coin, the head and tail will appear one time each.

If you toss a coin two times, then there is one chance that two heads will appear, two chances of one head and one tail appearing, and one chance of getting two tails showing the coefficient of the 3rd row—121.

Similarly, if you toss a coin three times, there is only a single chance to get three heads, three chances of getting two heads and one tail, three chances of getting one head and two tails and one for all tails revealing the pattern, 1331.

Tossing a Coin—No. of Times	Possible Results	Coefficient of Rows
1	H	1
	T	1
2	HH	1
	HT, TH	2
	TH	1
3	HHH	1
	HHT, HTH, THH	3
	HTT, TTH, THT	3
	TTT	1

So do the magic of expanding algebraic identities and explore some more properties of the Pascal Triangle!

Practice Problems

Expand the following:

a) $(a + b)^4 + (a - b)^4$
b) $(\sqrt{2} + \sqrt{5})^6 - (\sqrt{2} - \sqrt{5})^6$
c) $(x + y)^5 + (x - y)^5$
d) $(2 + \sqrt{5})^3 - (2 - \sqrt{5})^3$
e) $(11\sqrt{2} + 2\sqrt{5})^4 - (11\sqrt{2} - 6\sqrt{5})^4$

Finding the Area of a Polygon

In geometry, we find the area of different two-dimensional objects such as the square, rectangle, triangle, rhombus, trapezium, parallelogram, etc.; likewise three-dimensional objects include the cube, cone, cuboid, frustum, cylinder, sphere, etc. Finding the area, lateral area or total surface area of any of these objects simply requires the use of formula and it becomes easy to handle such questions, but when it comes to find the area of the irregular polygon. In such cases, we generally use the graph to draw such a figure or there already exists an image in the book we come across the question in and we count the number of squares inside the polygon and estimate the area. Many a times, while dealing with such questions, we use the geo-board and calculate the area.

The method taught in class really disturbs me a lot as it comes with no formula, nor is it deductive. But I have a solution for such a problem which I generally use in my class and I get a positive reply from most of the students who use it.

It was an Austrian mathematician named George Alexander Pick (10 August 1859–26 July 1942) who came up with the solution to find the area of lattice polygons; the solution is better known as **Pick's theorem**.

Suppose you have a simple polygon constructed on a grid of equal distances as done on a geo-board onto which a certain number of nails are half-drilled or a polygon constructed on graph paper—in both circumstances, you will be left with no choice but to count the nails or the squares.

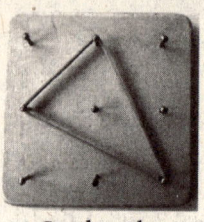
Geo-board

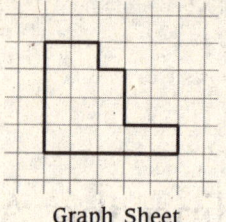
Graph Sheet

Now, the Pick's Theorem: For a given polygon constructed on a grid of equal distances, its area A is given by $A = i + b/2 - 1$, where i = interior point located in the polygon

b = lattice points on the boundary placed on polygon's perimeter

Let's take an example: Find the area of the polygon given below.

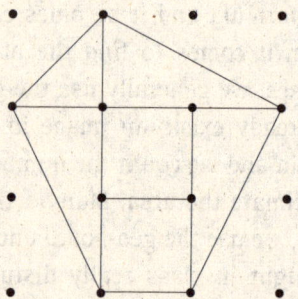

Solution: You can colour the diagram and count the number of squares inside the image. If the coloured area in any square is more than half, count that as 1 square unit.

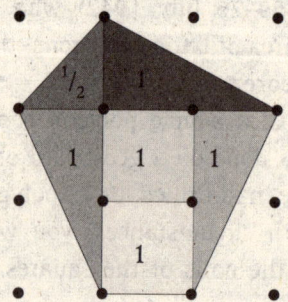

Hence, total area = 1 + 1 + 1 + 1 + 1 + $1/2$ = $5 \frac{1}{2}$ square unit

Now, if you know the Pick's theorem, then you need not colour the image in different shades or count the number of squares inside it keeping in mind whether the square inside any particular image is more than half or less than half of it. Let's now move to solving this question with the help of the Pick's theorem.

Here you can see a number of points inside the polygon = 4 = i

Number of points on the boundary = 5 = b

Hence, Area (A) = i + b/2 − 1
$$= 4 + 5/2 - 1$$
$$= 5 \frac{1}{2} \text{ square units}$$

Isn't this easy to calculate?

Example: Find the area of the given figure.

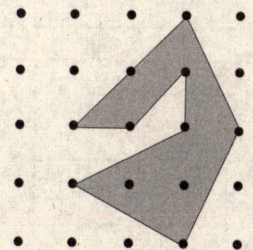

Solution:

Here, you can see number of points inside the polygon is 2 = i

Number of points on the boundary = 9 = b

Hence, Area (A) = i + b/2 − 1
$$= 2 + 4.5 - 1$$
$$= 5.5 \text{ square units}$$

Example: Find the area of the given figure.

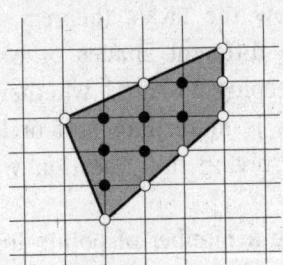

Solution: Interior point = 7
Boundary point = 8
Hence, Area (A) = $i + b/2 - 1$
$= 7 + 4 - 1$
$= 10$ square units

Example: Find the area of the following figure.

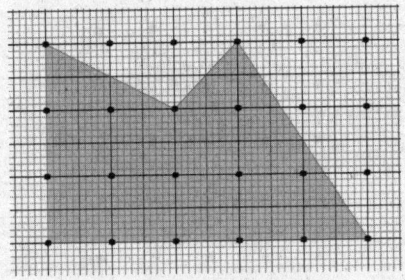

Solution: Interior point = 6
Boundary point = 11
Hence, Area (A) = $i + b/2 - 1$
$= 6 + 5.5 - 1$
$= 10.5$ square units

Example: Find the area of the given figure.

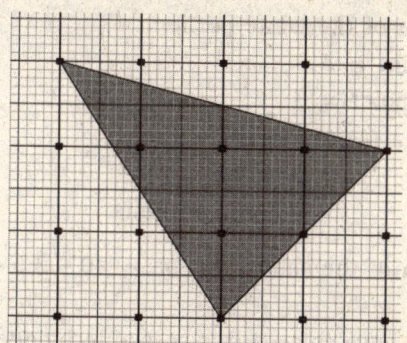

Solution: Interior point = 4
Boundary point = 4
Hence, Area (A) = i + b/2 − 1
= 4 + 2 − 1
= 5 square units

Example: Find the area of the shaded region.

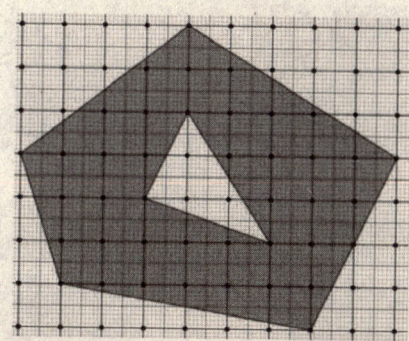

Solution: Here the area of the shaded region can be calculated by subtracting the area of the white portion from the whole polygon.

Area of the shaded region (including the white portion),
Interior point = 38
Boundary point = 6
Hence, Area (A) = i + b/2 − 1
= 38 + 3 − 1
= 40 square units

Again, area of the white portion,
Interior Point = 3
Boundary point = 3
Hence, Area (A) = i + b/2 − 1
= 3 + 1.5 − 1
= 3.5 square units

Therefore, area of the shaded region = 40 − 3.5 = 36.5 square units.

I hope you have enjoyed this special technique of finding the area of a polygon. This one is not only easy but quite interesting. Let's do some exercises.

Practice Problems

Find the area of the following polygons:

a)
b)
c)
d)

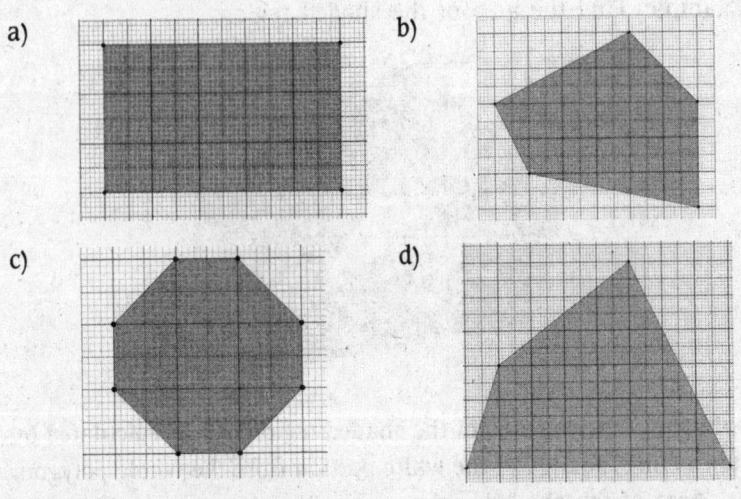

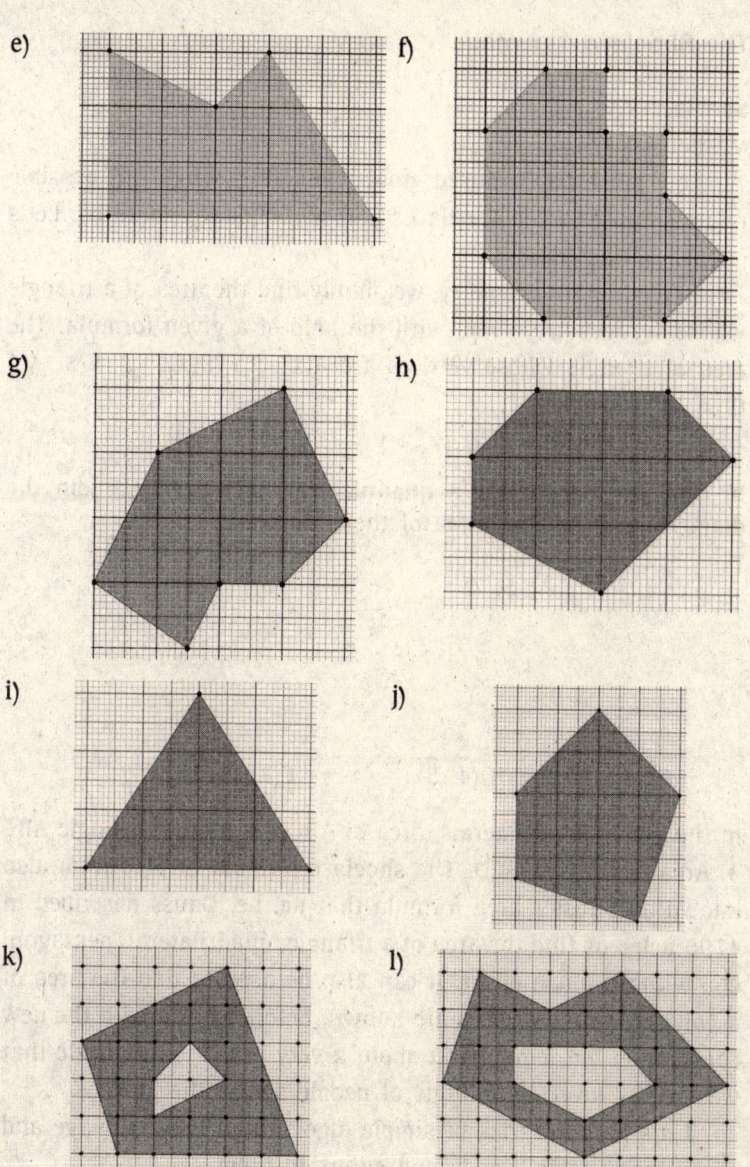

The Shoelace Algorithm

Shoelaces?
Mathematics?

You must be confused now wondering what the possible relationship between shoelaces and mathematics could be. Let's quench your curiosity.

In coordinate geometry, we simply find the area of a triangle whose vertices are given, with the help of a given formula. The area of triangle whose vertices are $A(x_1, y_1)$, $B(x_2, y_2)$, $C(x_3, y_3)$ then is

$$\frac{1}{2} [x_1 (y_2 - y_3) + x_2 (y_3 - y_1) + x_3 (y_1 - y_2)]$$

In case the polygon is a quadrilateral, then we first join the diagonal and find the area of the triangle.

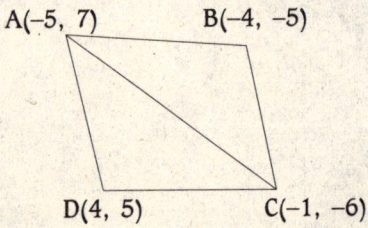

In the given quadrilaterals, area of ABCD = Area of triangle ABC + Area of triangle ACD. The shoelace formula or algorithm also known as Gauss's area formula that he, i.e. Gauss described in 1795 helps us find the area of a triangle, quadrilateral, pentagon, etc. without much effort. It can also be used to find the area of a polygon whose vertices are known. Before I begin with the new concept, let me remind you about a very sensational puzzle that was wildly loved by billions of people across the globe.

Let's play the trick of simple algebra and know the age and shoe size of your friends and entertain them.

Follow the simple steps and enjoy this puzzle:

- Take your shoe size. The shoe size should not be in halves.
- Multiply it by 5.

- Add 50 to the previous result.
- Multiply it by 20.
- Add 1016 (in year 2016), 1017 (in 2017), 1018 (in 2018).
- Subtract the year you were born in.

You will get a three-digit number in the end. The last number (number at hundreds' place) will give you the shoe size and the two remaining digits (digits at the ones' and tens' places) will tell your age.

Example: Now let's see these steps applied. Nishtha was born in 2015 and she wears a shoe size 4.

Shoe size = 4

Multiply it by 5, i.e. $4 \times 5 = 20$

Add 50; $20 + 50 = 70$

Multiply the previous result by 20; $20 \times 70 = 1400$

Add 1018; $1400 + 1018 = 2418$

Subtract the birth year; $2418 - 2015 = 4\ 03$

Here the digit at the hundreds' place = 4 will give you the shoe size

The remaining two digits, 03 = Present age

Use the concept of algebra to find the reason behind this puzzle.

Now let's focus on the shoelace theorem.

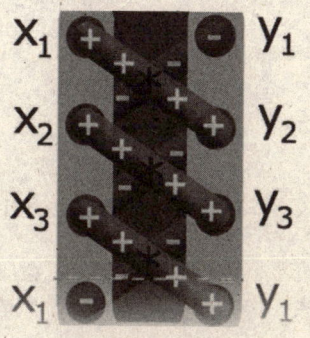

Finding the Area of a Polygon

In the left side of the image the shoelace has been shown with x-coordinates and y-coordinates whereas in the right side you can see the original shoe with the lace knotted accordingly.

How to name coordinates on shoelace

First, take the coordinates of a triangle. In the triangle, we have three vertices, namely: (x_1, y_1), (x_2, y_2) and (x_3, y_3). On the left side of the shoelace structure, write the first three x coordinates and on the right side, the y coordinates. At the bottom, write the first coordinates again as shown.

Area of triangle = Sum of cross-section of downward diagonals − Cross-section of upward diagonals

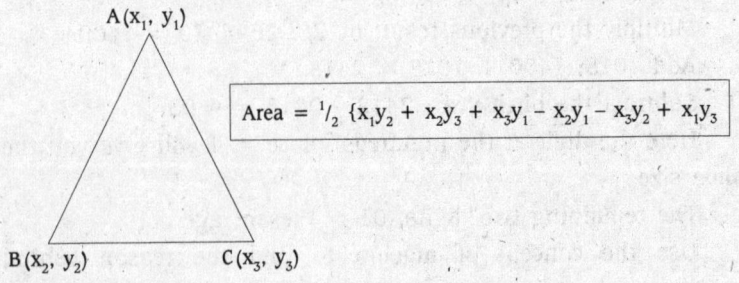

$$\text{Area} = \frac{1}{2} \{x_1y_2 + x_2y_3 + x_3y_1 - x_2y_1 - x_3y_2 + x_1y_3\}$$

Example: Find the area of a triangle whose vertices are (2, −3), (4, −1) and (0, 2).
Solution:

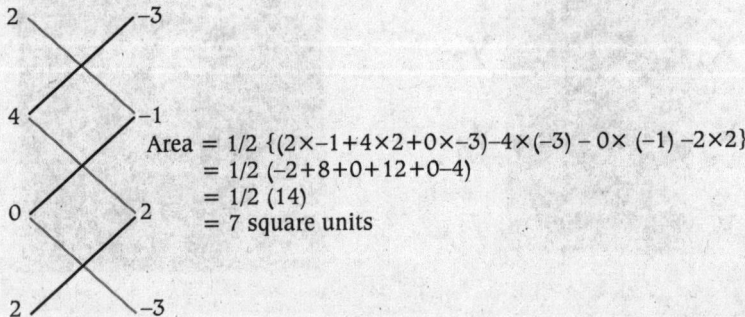

Area = 1/2 {(2×−1+4×2+0×−3)−4×(−3) − 0× (−1) −2×2}
= 1/2 (−2+8+0+12+0−4)
= 1/2 (14)
= 7 square units

Example: Find the area of a triangle whose vertices are A (15, 15), B (23, 30) and C (50, 25).
Solution:

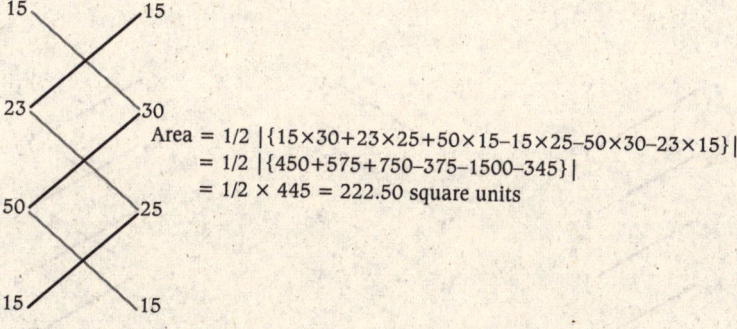

Area = 1/2 |{15×30+23×25+50×15−15×25−50×30−23×15}|
     = 1/2 |{450+575+750−375−1500−345}|
     = 1/2 × 445 = 222.50 square units

Area of a Quadrilateral

In a simple method, while finding the area of a quadrilateral, we simply join one of its opposite vertices, dividing the quadrilateral into two triangles and later find the area of each of the triangles and their sum finds the area of the quadrilateral.

Example: Find the area of the quadrilateral whose vertices are (5, 2), (5, −1), (−2, −1) and (−2, 2).
Solution:

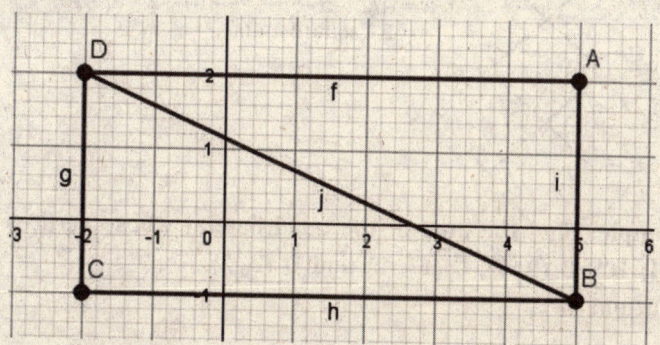

Since, ar (ABCD) = ar (ABD) + ar (BCD)

So area in this case can either be found using the triangle formula as described above or you can directly find the area of the quadrilateral as shown below.

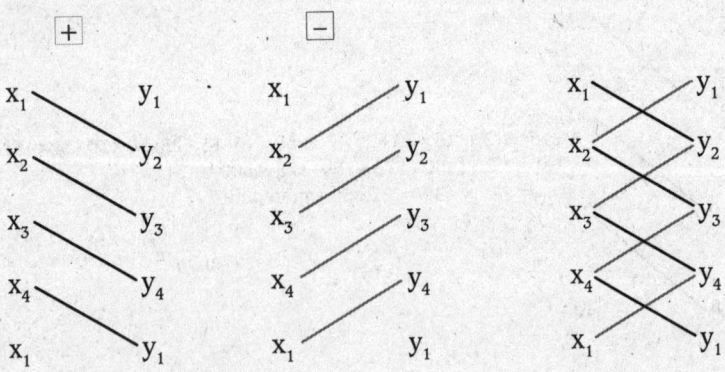

Area of quadrilateral =

$$= \frac{1}{2} |x_1 y_2 + x_2 y_3 + x_3 y_4 + x_4 y_1 - x_2 y_1 - x_3 y_2 - x_4 y_3 - x_1 y_4|$$

Area = 1/2 (−5 −5 −4 −4 −10 − 2 − 2 − 10)
     = 21 square units

Example: Find the area of the quadrilateral whose vertices are A (4, 10), B (9, 7), C (11, 2) and D (2, 2).
Solution:

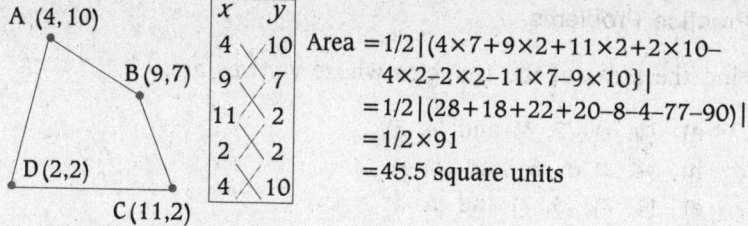

Area = 1/2|(4×7+9×2+11×2+2×10−
       4×2−2×2−11×7−9×10)|
     = 1/2|(28+18+22+20−8−4−77−90)|
     = 1/2×91
     = 45.5 square units

Example: Find the area of the given polygon.

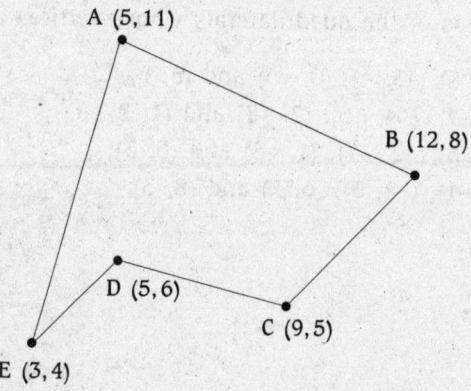

Solution:

Area of Pentagon

$$= \frac{1}{2}|x_1y_2 + x_2y_3 + x_3y_4 + x_4y_5 + x_5y_1 - x_2y_1 - x_3y_2 - x_4y_3 - x_5y_4 - x_1y_5|$$

$$A = \frac{1}{2}|3 \times 11 + 5 \times 8 + 12 \times 5 + 9 \times 6 + 5 \times 4$$
$$- 4 \times 5 - 11 \times 12 - 8 \times 9 - 5 \times 5 - 6 \times 3|$$

$$= \frac{60}{2} = 30 \text{ square units}$$

Finding the area of a triangle or quadrilateral with given vertices in a coordinate plane, can be easily calculated with the help of the shoelace method/algorithm. You don't need to memorize the formula as it is self-explanatory. So keep calculating the area of a two-dimensional figure when the coordinate of each corner is given and enjoy the shoelace method.

Finding the Area of a Polygon

Practice Problems

Find the areas of the triangles whose vertices are:

a) (1, 1), (2, 3) and (4, 5)
b) (4, 2) (6, 5) and (1, 4)
c) (1, 2), (3, 7) and (5, 3)
d) (0, −1), (2, 1) and (0, 3)
e) (−2, 0), (2, 0) and (0, 2)

Find the areas of the quadrilaterals whose vertices are:

a) (−4, 8) (−3, 4), (0, −5) and (5, 6)
b) (5, −3), (−4, −6), (2, −3) and (1, 2)
c) (−3, 5) (−2, −7), (1, −8) and (6, 3)
d) (0, −1), (−2, 3), (6, 7) and (8, 3)

VI
SQUARES AND CUBES ARE MUCH FUN

Square

Square refers to multiplying a number twice. If 'a' is a number then $a \times a = a^2$ is called the square of 'a'.[*]

Finding the Square of a Number That Ends With 5

Squaring a number ending with 5 is very simple.

- Square the unit digit 5 and write 25 at the end.
- The remaining digit should be multiplied with the next digit and has to be placed before 25.

$$(A5)^2 = A \times (A+1) / 25$$

Example: Find the square of 25.
Solution: $(25)^2 = 2 \times 3/25$
$ = 625$

Example: Find the square of 55.
Solution: $(55)^2 = 5 \times 6/25$
$ = 3025$

---

[*]Of the methods to be discussed in this book, I won't be including the Vedic method and the Trachtenberg system since these have been covered in my previous publications, *The Essentials of Vedic Mathematics* and *Maths Made Easy*.

Example: Find the square of 85.
Solution: $(85)^2 = 8 \times 9/25$
$= 7225$

Example: Find the square of 115.
Solution: $(115)^2 = 11 \times 12/25$
$= 13225$

Squaring a Number Ending With 25

Squaring a number ending with 25 is as easy as squaring a number ending with 5 as discussed above. The whole operation can be done in two steps:

1. Write 25 at the end.
2. Add 3 to the number before 5 and multiply it with the number (before 5).

$$(a25)^2 = a \times (a2 + 3) / 625$$
$$(ab25)^2 = ab \times (ab2 + 3) / 625$$

Example: Find the square of 125.
Solution: $(125)^2 = 1 \times (12 + 3) / 625$
$= 15 / 625$
$= 15625$

Example: Find the square of 625.
Solution: $(625)^2 = 6 \times (62 + 3) / 625$
$= 6 \times 65 / 625$
$= 390 / 625$
$= 390625$

Example: Find the square of 1225.
Solution: $(1225)^2 = 12 \times (122 + 3) / 625$
$= 12 \times 125 / 625$
$= 1500 / 625$
$= 1500625$

Example: Find the square of 1625.
Solution: $(1625)^2 = 16 \times (162 + 3) / 625$
$= 16 \times 165 / 625$
$= 2640 / 625$
$= 2640625$

Example: Find the square of 1825.
Solution: $(1825)^2 = 18 \times (182 + 3) / 625$
$= 18 \times 185 / 625$
$= 3330 / 625$
$= 3330625$

Squaring a Number That Has 1 Repeating

Let's first check the pattern of such a squaring.
$11 \times 11 = 121$
$111 \times 111 = 12321$
$1111 \times 1111 = 1234321$
$11111 \times 11111 = 123454321$

If you observe the above pattern of squaring minutely, you can find the following points:

a) Count the number of repeated 1s; say it is n.
b) Write 1 to 'n' in ascending order.
c) After you reach n, start writing them in decreasing order up to 1.

Now, let's find the square of 1111111.

Here, number of repeated 1s is 7. Start writing 1 to 7 and then move in decreasing order.
$(1111111)^2 = 1234567654321$
$(11111111)^2 = 123456787654321$

Squaring a Number That Has Two Digits

The method to find the square of any two-digit number is also very simple and is based on an algebraic identity.

First, make three groups.

I	II	III
Square of ten's digit	1st digit × 2nd digit	Square of 2nd digit

Write the result of the second part once more.

I	II	III
Square of ten's digit	1st digit × 2nd digit	Square of 2nd digit
	1st digit × 2nd digit	

Add each column, beginning from the right side. Only one digit should be kept in the second and third columns respectively. If there is more than one digit in these columns, transfer the leftmost digit to the previous column.

Example: Find the square of 13.
Solution:

1	3 +3	9
1	6	9

Example: Find the square of 29.
Solution:

2 × 2 = 4	2 × 9 = 18 + 18	9 × 9 = 81
4	36	81

= 4   36 + 8 = 44   1
= 4 + 4   4   1
= 841

Example: Find the square of 67.
Solution:

| 6 × 6 = 36 | 6 × 7 = 42
+42 | 7 × 7 = 49 |

     36   84   49
= 36   84 + 4 = 88   9
= 36 + 8   8   9
= 4489

Example: Find the square of 93.
Solution:

| 9 × 9 = 81 | 9 × 3 = 27
+27 | 3 × 3 = 9 |

     81   54   9
= 81 + 5 = 86   4   9
= 8649

Method of Squaring a Number Near Base 10, 100, 1000, 1000, etc.

You can find the square of any number that is near 10, 100, 1000, etc., in a step or two. Just follow the instructions and gain mastery over it.

- Find the difference of number with base number that is 10, 100, 1000, etc.
- If the difference is positive, add it to the number to be squared. If the difference is negative, get it subtracted from the number.
- Write the square of the difference after the result obtained

in the second step. The number of digits in the square of the difference will be decided by the number of zeros in the base. If the number of zeros in the base is 2, there will be two digits on the right side. If the number of digits is more than 2, transfer the leftmost digit to the left side. In case the number of digits is less than the prescribed number of digits, put zero before it so as to equate it to the number of zeros in the base. The number of digits on the right side depends upon the number of zeros in base number 10, 100, etc.

Square of a number = Number + Difference / Difference square

Example: Find the square of 12.
Solution: Base = 10
Difference = 12 − 10 = + 2
$(12)^2 = 12 + 2 / 2^2$
    = 144

Example: Find the square of 17.
Solution: Base = 10
Difference = 17 − 10 = + 7
$(17)^2 = 17 + 7 / 7^2$
    = 24 / 49
    = 24 + 4 / 9 (Since base 10 has 1 zero, so the leftmost digit 4 of 49 is transferred and added to 24.
    = 289

Example: Find the square of 92.
Solution: Base = 100
Difference = 92 − 100 = −8
$(92)^2 = 92 − 8 / 8^2$
    = 8464

Example: Find the square of 97.
Solution: Base = 100

Difference = 97 − 100 = −3
$(97)^2$ = 97 − 3 / 3^2
    = 94 / 9 (Since base 100 has 2 zeros, 1 zero will have to be placed before 9)
    = 94/09

Example: Find the square of 99.
Solution: Base = 100
Difference = 99 − 100 = −1
$(99)^2$ = 99 − 1 / 1^2
    = 98/1 (Since base 100 has 2 zeros, 1 zero will have to be placed before 1)
    = 98/01

Example: Find the square of 102.
Solution: Base = 100
Difference = 102 − 100 = + 2
$(102)^2$ = 102 + 2 / 2^2
    = 104/04 (Since base 100 has 2 zeros, 1 zero will have to be placed before 1)

Example: Find the square of 108.
Solution: Base = 100
Difference = 108 − 100 = + 8
$(108)^2$ = 108 + 8 / 8^2
    = 116/64

Example: Find the square of 992.
Solution: Base = 1000
Difference = 992 − 1000 = −8
$(992)^2$ = 992 − 8 / 8^2 (Since base 1000 has 3 zeros, 1 zero will have to be placed before 64)
    = 984064

Example: Find the square of 997.
Solution: Base = 1000
Difference = 997 − 1000 = −3

$(997)^2 = 997 - 3 / 3^2$

$ = 994 / 9$ (Since base 1000 has 3 zeros, 2 zeros will have to be placed before 9)

$ = 994/009$

Example: Find the square of 10002.
Solution: Base = 10000

Difference = 10002 − 10000 = + 2

$(10002)^2 = 10002 + 2 / 2^2$

$ = 10004/0004$ (Since base 10000 has 4 zeros, 3 zeros will have to be placed before 1)

You can extend it likewise.

Square of a Number that is Near to Base 50, 500, 5000, etc.

Squaring of a number that is near base 50 (41–60), 500 (475–525), 5000, etc., can be done with a small change in the method mentioned above.

- Find the difference of the number to be squared from 50 and place it on the left side.
- Divide the number placed on the left side by 2.
- Place the square of difference from 50 on the right side. Number placed on the right side should be equal to the number of zeros in base + 1. If the number to be squared is near base 50, there should be 2 digits on the right side. If it is near base 500, then there should be 3 digits on the right side. In case the number of digits on the right side is less than the desired digit, put zero(s) as discussed in the above method and if the digits on the right side is more, add the extreme left hand side digit of right side to the left side of the number.

Example: Find the square of 47.
Solution: Base = 50

Difference = 47 − 50 = −3

$(47)^2 = 47 - 3 / (-3)^2$
$= 44 / 9$

Divide the left part by 2
$= {}^1/_2 \times 44 / 09$
$= 2209$ (For base 50, number of digits on the right side should be 2)

Example: Find the square of 54.
Solution: Base = 50
Difference = 54 − 50 = 4
$(54)^2 = 54 + 4 / (4)^2$
$= 58 / 16$

Divide the left part by 2
$= {}^1/_2 \times 58 / 16$
$= 2916$

Example: Find the square of 67.
Solution: Base = 50
Difference = 67 − 50 = 17
$(67)^2 = 67 + 17 / (17)^2$
$= 84 / 289$

Divide the left part by 2
$= {}^1/_2 \times 84 / 289$
$= 42 + 2 / 89$ (For base 50, number of digits on the right side should be 2)
$= 4489$

Example: Find the square of 497.
Solution: Base = 500
Difference = 497 − 500 = −3
$(497)^2 = 497 - 3 / (-3)^2$
$= 494 / 9$

Divide the left part by 2
$= {}^1/_2 \times 494 / 009$
$= 247009$ (For base 500, number of digits on the right side should be 3)

If you have learnt by heart the square of numbers up to 25, you can find the square up to 75 in a few simple steps and that too, in your mind, without having to use pen and paper. The method of squaring near base will come to your help when calculating squares up to 20. Let's first memorize the chart given here.

$11^2 = 121$	$12^2 = 144$	$13^2 = 169$	$14^2 = 196$	$15^2 = 225$
$16^2 = 256$	$17^2 = 289$	$18^2 = 324$	$19^2 = 361$	$20^2 = 400$
$21^2 = 441$	$22^2 = 484$	$23^2 = 529$	$24^2 = 576$	$25^2 = 625$

Let's see how to find the square of numbers beyond 25 with ease.

Square of a Number From 26 to 75

> Square of a number above 25 = 25 ± Difference from 50 + Square of difference

Example: Find the square of 29.
Solution: $(29)^2$
= 25 ± (29 − 50) / 21 × 21
= 25 − 21 / 441
= 4 + 4 / 41
= 841

Example: Find the square of 34.
Solution: $(34)^2$
= 25 ± (34 − 50) / 16 × 16
= 25 − 16 / 256
= 9 + 2 / 56
= 1156

Example: Find the square of 58.
Solution: $(58)^2$
= 25 ± (58 − 50) / 8 × 8
= 25 + 8 / 64

= 33 / 64
= 3364

Example: Find the square of 69.
Solution: $(69)^2$

= 25 ± (69 − 50) / 19 × 19
= 25 + 19 / 361
= 44 + 3/ 61
= 4761

Example: Find the square of 73.
Solution: $(73)^2$

= 25 ± (73 − 50) / 23 × 23
= 25 + 23 / 529
= 48 + 5/ 29
= 5329

Some Special Methods to Calculate Squares

Squaring a number that begins with 3 and ends with any single non-zero digit

We know $(a + b)^2 = a^2 + 2ab + b^2$
Let's elaborate the formula.

$$(a + b)^2 = a^2 + 2ab + b^2$$
$$= a(a + 2b) + b^2$$
$$= a(a + b + b) + b^2$$
$$= a(n + b) + b^2 \text{ where } n = a + b$$

Now, let's use this in finding the square of 31 to 38 and then we shall explore it with 3 repeated n times and the last digit ending with a non-zero number.

$(30 + b)^2 = 30(n + b) + b^2$, where $n = 30 + b$ (check the above expansion)

Using this formula, we can find the square of 31 to 38. See below:

$31^2 = 30(31 + 1) + 1^2 = 960 + 1 = 961$
$32^2 = 30(32 + 2) + 2^2 = 1020 + 4 = 1024$
$33^2 = 30(33 + 3) + 3^2 = 1080 + 9 = 1089$
$34^2 = 30(34 + 4) + 4^2 = 1140 + 16 = 1156$
$35^2 = 30(35 + 5) + 5^2 = 1200 + 25 = 1225$
$36^2 = 30(36 + 6) + 6^2 = 1260 + 36 = 1296$
$37^2 = 30(37 + 7) + 7^2 = 1320 + 49 = 1369$
$38^2 = 30(38 + 8) + 8^2 = 1380 + 64 = 1444$

Here, we will divide the above group into two parts—

a) 31, 34, 37 in the first group, where the difference between the numbers is 3.
b) 32, 35, 38 in the second group where the difference between the numbers is 3.

Let's first begin with the first group. For a while, take the first number 31.

$31^2 = \mathbf{0961}$

We will take 09 and 61 separately.

Example: Find the square of 331.
Solution: Here 3 is repeated two times. While squaring, 1 and 5 will be repeated one less than the recurring of 3. First write 09 and 61 with a gap.

$(331)^2 = \ldots\ldots\ldots\ 09\ \ldots\ldots\ldots\ 61$

Fill 1 before 09 and 5 before 61. The number of times 1 will be written depends upon the number of times 3 appears in the number to be squared. If 3 is repeated 5 times, 1 and 5 will be repeated four times. Hence,

$(331)^2 = 1\ \mathbf{09}\ 5\ \mathbf{61}$

Example: Find the square of 3331.
Solution: Here 3 is repeated three times. Obviously while squaring, 1 and 5 will be repeated two times.

$(3331)^2$ = ... 09 61
= 11 09 55 61

Now, without wasting time, let's find the square instantly.

$(33331)^2$ = ... 09 61
= 111 **09** 555 **61**

$(333331)^2$ = ... 09 61
= 1111 **09** 5555 **61**

$(3333331)^2$ = ... 09 61
= 11111 **09** 55555 **61**

$(33333331)^2$ = ... 09 61
= 111111 **09** 555555 **61**

Now, let's find the square of a number that has 3 recurring and ends with 4.

$(34)^2 = 1156$

We will take 11 and 56 separately.

Example: Find the square of 334.

Solution: Here 3 is repeated two times. While squaring, 1 and 5 will be repeated one less than the recurring of 3. First write 11 and 56 with a gap.

$(334)^2$ = ... 11 56

Fill 1 before 11 and 5 before 56. The number of times 1 will be written, depends upon the number of times 3 appears in the number to be squared. If 3 is repeated five times, 1 and 5 will be repeated four times. Hence,

$(334)^2$ = 1 **11** 5 **56**

Example: Find the square of 3334.

Solution: Here 3 is repeated three times. Obviously, while squaring, 1 and 5 will be repeated two times. First write 11 and 56 with a gap.

$(3334)^2$ = ...11 56

Fill 1 before 11 and 5 before 56. The number of times 1 will be written, depends upon the number of times 3 will appear in the number to be squared. If 3 is repeated five times, 1 and 5 will be repeated four times. Hence,

$(3334)^2 = 11\ 11\ 55\ 56$

Let's see some more examples.

$(33334)^2$ = ...11 56
= 111 11 555 56
$(333334)^2$ = ...11 56
= 1111 11 5555 56
$(3333334)^2$ = ...11 56
= 11111 11 55555 56
$(33333334)^2$ = ...11 56
= 111111 11 555555 56

Now, let's find the square of a number that has 3 recurring and ends with 7.

$(37)^2 = 1369$

We will take 13 and 69 separately.

Example: Find the square of 337.
Solution: Here 3 is repeated two times. While squaring, 1 and 5 will be repeated one less than the recurring of 3. First write 13 and 69 with a gap.

$(337)^2$ = ... 13 69

Fill 1 before 13 and 5 before 69. The number of times 1 will be written, depends upon the number of times 3 appears in the number to be squared. If 3 is repeated five times, 1 and 5 will be repeated four times. Hence,

$(337)^2 = 1\ 13\ 5\ 69$

Example: Find the square of 3337.
Solution: Here 3 is repeated three times. Obviously while squaring, 1 and 5 will be repeated two times. First write 13 and 69 with a gap.

$(3337)^2$ = ...13 69

Fill 1 before 13 and 5 before 69. The number of times 1 will be written, depends upon the number of times 3 appears in the number to be squared. If 3 is repeated five times, 1 and 5 will be repeated four times. Hence,

$(3337)^2 = 11\ 13\ 55\ 69$

Let's do some more examples.

$(33337)^2 = $ 13 69
$= 111\ 13\ 555\ 69$
$(333337)^2 = $ 13 69
$= 1111\ 13\ 5555\ 69$
$(3333337)^2 = $ 13 69
$= 11111\ 13\ 55555\ 69$
$(33333337)^2 = $ 13 69
$= 111111\ 13\ 555555\ 69$

Now, let's begin the squaring of the second group. Here, the number ends with 2, 5 or 8 with recurring of the digit preceding it.

$32^2 = 1024$

We will take 10 and 24 in separately.

Example: Find the square of 332.

Solution: Here 3 is repeated two times. While squaring, 1 and 2 will be repeated one less than the recurring of 3. First write 10 and 24 with a gap.

$(332)^2 = $ 10 24

Fill 1 before 10 and 2 before 24. The number of times 1 will be written, depends upon the number of times 3 appears in the number to be squared. If 3 is repeated five times, 1 and 2 will be repeated four times. Hence,

$(332)^2 = 1\ 10\ 2\ 24$

Example: Find the square of 3332.

Solution: Here 3 is repeated three times. While squaring, 1 and 2 will be repeated one less than the recurring of 3. First write 10 and 24 with a gap.

$(3332)^2 = $ 10 24

Fill two times 1 before 10 and two times 2 before 24. The number of times 1 will be written, depends upon the number of times 3 comes in the number to be squared.

Hence,

$(3332)^2 = 11\ \mathbf{10}\ 22\ \mathbf{24}$

Let's do some more examples.

$$\begin{aligned}
(33332)^2 &= \text{..........}10\text{ ....}24 \\
&= 111\ \mathbf{10}\ 222\ \mathbf{24} \\
(333332)^2 &= \text{..........}10\text{ ....}24 \\
&= 1111\ \mathbf{10}\ 2222\ \mathbf{24} \\
(3333332)^2 &= \text{..........}10\text{ ....}24 \\
&= 11111\ \mathbf{10}\ 22222\ \mathbf{24} \\
(33333332)^2 &= \text{..........}10\text{ ....}24 \\
&= 111111\ \mathbf{10}\ 222222\ \mathbf{24}
\end{aligned}$$

Square of a number ending with 9

$$\begin{aligned}
(a + 9)^2 &= (a + 10)^2 - (a + 9 + a + 10) \\
&= a^2 + 20a + 100 - 2a - 19 \\
&= a^2 + 18a + 81
\end{aligned}$$

Example: $(19)^2 = (20)^2 - (19 + 20)$
$= 400 - 39 = 361$

Example: $(49)^2 = (50)^2 - (49 + 50)$
$= 2500 - 99$
$= 2401$

Example: $(89)^2 = (90)^2 - (89 + 90)$
$= 8100 - 179$
$= 7921$

Example: $(99)^2 = (100)^2 - (99 + 100)$
$= 10000 - 199$
$= 9801$

Squaring a number that begins with 6 and ends with 4

Let's begin with 64.

$(64)^2 = 4096$

Now we shall deal with squaring a number that has 6 recurring and ends with 4.

$(664)^2 = ?$
$(6664)^2 = ?$
$(66664)^2 = ?$
$(66666666664)^2 = ?$

Rule: As we have seen that $(64)^2 = 4096$, here we shall take 40 and 96 as two groups.

Write 40 and 96 separately.

.... 40 96

Before 40, write 4 repeated n – 1 times

Then write 8 repeated n – 1 times after 40 and before 96.

Example: Find the square of 664.
Solution: Here 6 is repeated two times.

First write 40 and 96 with a gap before and after.

............40 96

Since 6 is repeated two times, write 4 and 8, once in between.

Hence, $(664)^2 = $ 4 40 8 96

Example: Find the square of 6664.
Solution: Here 6 is repeated three times.

First write 40 and 96 with a gap before and after.

............40 96

Since 6 is repeated three times, write 4 and 8, 3 – 1 = two times in between.

Hence, $(664)^2 = $ 44 40 88 96

Similarly, we can find the square of the following numbers mentally.

a) $(66664)^2 = $ 444 40 888 96

b) $(666664)^2 = 4444\ 40\ 8888\ 96$
c) $(6666664)^2 = 44444\ 40\ 88888\ 96$

Squaring a number that has 6 recurring and ends with 7

Let's begin with 67.
$(67)^2 = 4489$
Now we shall deal with squaring a number that has 6 repeated and ends with 7.

$(667)^2 = ?$
$(6667)^2 = ?$
$(66667)^2 = ?$
$(66666666667)^2 = ?$

Rule: As we have seen that $(67)^2 = 4489$, here we shall take 44 and 89 as two groups.

Write 44 and 89 separately.

...........44 89

Before 44, write 4 repeated n – 1 times.

Then write 8 repeated n – 1 times after 44 and before 89.

Example: Find the square of 667.
Solution: First write 44 and 89 separately.

...........44 89

In 667, 6 is repeated two times.

Before 44, write 4 repeated n – 1 times.
Then write 8 repeated n – 1 times after 44 and before 89.
Hence, $(667)^2 = 4\ 44\ 8\ 89$

Example: Find the square of 66667.
Solution: First write 44 and 89 separately.

...........44 89

In 66667, 6 is repeated four times.

Before 44 write 4 repeated n – 1 times, i.e. three times.

Then write 8 repeated n – 1 times after 44 and before 89.

Hence, $(667)^2 = 4\ 44\ 8\ 89$

Example: Find the square of 66666667.
Solution: First write 44 and 89 separately.
............44 89
In 66666667, 6 is repeated seven times.
Before 44 write 4 repeated six times.
Then write 8 repeated six times after 44 and before 89.
Hence, $(66666667)^2 = 444444\ 44\ 888888\ 89$
Let's take another interesting case.

Squaring a number that has 6 recurring and ends with 2

Before coming to the general method, let's first find the square of 62.
   $(62)^2 = 3844$
Here, the two groups of numbers 38 and 44 need to be remembered separately. This will work as the key to squaring a number that has 6 repeating and ends with 2.
   Check out the following:

$(662)^2 = ?$
$(6662)^2 = ?$
$(66662)^2 = ?$
$(666662)^2 = ?$

Rule: Write 38 and 44 separately.
............38 44
Before 38 write 2 repeated n – 1 times.
Then write 2 repeated n – 1 times after 38 and before 44.
Write 4 as many times as the number of 2s and place it before 38.
   Now let's look at the following examples.

Example: Find the square of 662.
Solution: Write 38 and 44 separately.
............38 44

Here, 6 is repeated two times (n = 2).
Before 38 write 2 repeated n − 1 times.
Then write 2 repeated n − 1 times after 38 and before 44.
Hence, $(662)^2 = 4\ 38\ 2\ 44$

Example: Find the square of 66662.
Solution: Write 38 and 44 separately.

..........38 44

Here, 6 is repeated four times (n = 4).
Before 38 write 2 repeated 4 − 1 = 3 times.
Then write 2 repeated 4 − 1 = 3 times after 38 and before 44.
Hence, $(66662)^2 = 444\ 38\ 222\ 44$

Example: Find the square of 66666662.
Solution: Write 38 and 44 separately.

..........38 44

Here, 6 is repeated seven times (n = 7).
Before 38 write 2 repeated 7 − 1 = 6 times.
Then write 2 repeated 7 − 1 = 6 times after 38 and before 44.
Hence, $(66666662)^2 = 444444\ 38\ 222222\ 44$

Interesting Properties of Square Numbers

Here are some interesting properties of square numbers.

Square numbers are 1, 4, 9, 16, 25, 36, etc.

The Greeks also discovered that if consecutive odd numbers are added, they become square numbers. Like, $1 = 1 \times 1 = 1^2$

$1 + 3 = 4 = 2^2$
$1 + 3 + 5 = 9 = 3^2$
$1 + 3 + 5 + 7 = 16 = 4^2$
$1 + 3 + 5 + 7 + 9 = 25 = 5^2$

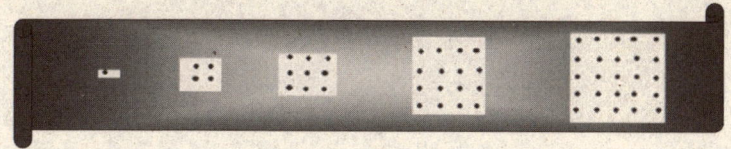

More interestingly, each higher square number is formed by adding an L-shaped set of pebbles to the previous number. The L-shape was called gnomon by the Greeks, which referred to an instrument imported to Greece from Babylon for measuring time.

Note that square numbers can be found by adding all triangular numbers in the following manner—

	1	3	6	10	15	21	28	36...
1	3	6	10	15	21	28	38...	
1	4	9	16	25	36	49	64...	

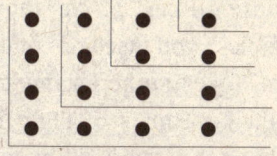

The L-shape figure above represents the pebbles in triangular numbers and when they are added to the subsequent triangular numbers, it gives the square numbers.

Practice Problems

Find the squares of the following by using the appropriate method:

a) 15
b) 25
c) 35
d) 65
e) 85
f) 44
g) 107
h) 112
i) 125
j) 76

Square Roots

In competitive examinations, there are two types of questions in the context of which you are asked to find square roots. In questions related to quadratic equations one needs to find $\sqrt{D}$ (where D stands for discriminant) = $b^2 - 4ac$. In the second type, you need to find the square root in mensuration problems of finding the area. For instance, find the area of an equilateral triangle whose side is 4 cm. Or find the area of a triangle whose sides are 10 cm, 14 cm and 16 cm respectively.

In both the examples, you need to find the square root of the irrational numbers. Take another example. The product of two numbers is 24 and their mean is 5. Find the numbers. In this example, you can either solve the quadratic equation $x^2 - 10x + 24 = 0$ by the middle term factorization method or by using the quadratic formula.

$$x = \frac{-b \pm \sqrt{b^2 - 4ac}}{2a}$$

Now, let's begin with explaining how to find the square root of a number that is not a perfect square.

Methods to Find the Square Root Numbers that are Not Perfect Squares

First method

The first method is very simple and it also takes little time when applied. In fact, with good practice you can get to solving relevant square root problems mentally. This is basically an estimation method that will give you the best approximate value, but for that, you need to learn the squares up to 20 or the following square root table.

Table 1

$\sqrt{1} = 1$	$\sqrt{36} = 6$	$\sqrt{121} = 11$	$\sqrt{256} = 16$
$\sqrt{4} = 2$	$\sqrt{49} = 7$	$\sqrt{144} = 12$	$\sqrt{289} = 17$
$\sqrt{9} = 3$	$\sqrt{64} = 8$	$\sqrt{169} = 13$	$\sqrt{324} = 18$
$\sqrt{16} = 4$	$\sqrt{81} = 9$	$\sqrt{196} = 14$	$\sqrt{361} = 19$
$\sqrt{25} = 5$	$\sqrt{100} = 10$	$\sqrt{225} = 15$	$\sqrt{400} = 20$

Here is the **approximation formula** to find the square root of a number.

Square root of Irrational Number =

$$\sqrt{\text{Nearest Perfect Square} + \frac{\text{Deviation from Irrational Number}}{2 \times \sqrt{\text{Nearest Perfect Square}}}}$$

Here, deviation is taken from the nearest square root. It may be more than or less than the nearest square root and the sign should be used accordingly.

Example: Find the square root of 23.
Solution: Perfect square approaching 23 is 25.

Deviation = 23 − 25 = −2

$$\sqrt{23} = \sqrt{25} - \frac{2}{2 \times \sqrt{25}}$$

$$= 5 - 1/5 = 5 - 0.2 = 4.8$$

Example: Find the square root of 123.
Solution: Perfect square approaching 123 is 121.

Deviation = 123 − 121 = +2

$$\sqrt{123} = \sqrt{121} + \frac{2}{2 \times \sqrt{121}}$$

$$= 11 + 1/11 = 11 + 0.09 = 11.09$$

Example: Find the square root of 231.
Solution: Perfect square approaching 231 is 225.

Deviation = 231 − 225 = +6

$$\sqrt{231} = \sqrt{225} + \frac{6}{2 \times \sqrt{225}}$$

$$= 15 + 1/5 = 15 + 0.2 = 15.02$$

Example: Find the square root of 602.
Solution: Perfect square approaching 602 is 625.

Deviation = 602 − 625 = −23

$$\sqrt{602} = \sqrt{625} - \frac{23}{2 \times \sqrt{625}}$$

$$= 25 - 23/50 = 25 - 11.50 = 13.50$$

There is another estimation method to find the square root of numbers that are not perfect squares.

Method 2

- First find the two numbers such that the number whose square root you are trying to find out is between them.
- Divide your number by one of those square roots.
- Take the average of the numbers chosen and the result obtained in the second step.
- Use the result of step 3 to repeat step 2 and 3 for a more accurate result.

Example: Find the square root of 13 up to 2 decimal places.
Solution: We know
$$3^2 < 13 < 4^2$$

- Divide 13 by 3: $13 \div 3 = 4.33$
- Average of 3 and 4.33: $(3 + 4.33)/2 = 3.66$
- Divide 13 by 3.66: $13 \div 3.66 = 3.55$
- Average of 3.66 and 3.55 = 3.60

On two repeated steps we can conclude that $\sqrt{13} = 3.60$

Example: Find the square root of 401 up to 2 decimal places.
Solution: We know
$$20^2 < 401 < 21^2$$

- Divide 401 by 20: $401 \div 20 = 20.05$
- Average of 20 and 20.05: $(20 + 20.05)/2 = 20.25$
- Divide 401 by 20.25: $401 \div 20.25 = 19.80$
- Average of 20.25 and 19.80 = 20.02

On two repeated steps we can conclude that $\sqrt{401} = 20.02$

Example: Find the square root of 701 up to 2 decimal places.
Solution: We know
$$26^2 < 701 < 27^2$$

- Divide 701 by 26: $701 \div 26 = 26.96$
- Average of 26 and 26.96: $(26 + 26.96)/2 = 26.48$
- Divide 701 by 26.48: $701 \div 26.48 = 26.47$
- Average of 26.48 and 26.47 = 26.47

On two repeated steps we can conclude that $\sqrt{701} = 26.47$

Method to Find Square Roots of Three-four-digit Numbers Mentally

So far we have learnt how to find the square root of any number that is not a perfect square. Now let's focus on a method that is simple to learn but it is effective only for numbers that are three to four digit long. Minutely observe this table that has the square of the first 10 natural numbers.

Table 2

N	N^2	Last Digit of N^2	Digit Sum of Numbers
1	1	1	1
2	4	4	4
3	9	9	9
4	16	6	7
5	25	5	7
6	36	6	9
7	49	9	4
8	64	4	1
9	81	1	9
10	100	00	1

Note:

- Squares of 1 and 9 end with 1 respectively.
- Squares of 2 and 8 end with 4 respectively.
- Squares of 3 and 7 end with 9 respectively.
- Squares of 4 and 6 end with 6 respectively.
- Square of 5 ends with 5.

Rule:

- Ignore the last two digits of the group and find the value of *a* from the above table.
- Obtain the last two digits of the square root from the second table. If the last digit of the square root to be extracted is 1, 4, 6, 9 then you will have two possibilities of answer, which will be determined in the next step.
- Multiply *a* by *a + 1*. If the product $a \times (a+1)$ > **first part of the group**, then the lower of the two possibilities will be taken.
- In case $a \times (a+1) \leq$ *first part of the group*, take the higher possible values for *b*.

Example: Find the square root of 2304.
Solution: First, group the numbers in a pair.

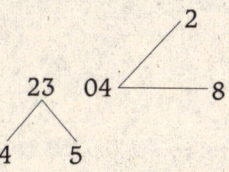

Ignore the last pair and find the value of **a** from Table 2.
$$4^2 < 23 < 5^2$$
Here, a = 4

Since the last digit of the number is 4, we will have two possible answers for **b**. It is either 2 or 8.

Multiply a = 4 by a + 1 = 5
$$4 \times 5 = 20 < 23 \text{ (First pair of group)}$$
Hence b = 8 (Higher of two possible values)
$\sqrt{2304} = 48$

Example: Find the square root of 1296.
Solution: We have two groups here—12 and 96. Ignore the last group (96) and find the value of *a* for the first group.

$3^2 < 12 < 4^2$

Hence, a = 3

As you can see, the last digit of 1296, that is the unit digit of second group 96, is 6. So from Table 1, we get two possible sets of answers for **b**, i.e. 4 and 6.

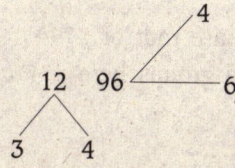

Multiply a = 3 by a + 1 = 4
$$3 \times 4 = 12 \text{ (First pair of group)}$$
Hence b = 6 (Higher of two possible values)
Therefore, $\sqrt{1296} = 36$

Example: Find the square root of 7744.
Solution: We have two groups here—77 and 44.

$8^2 < 77 < 9^2$

Hence, a = 8

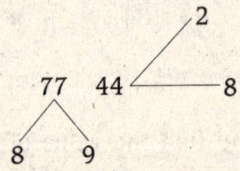

As you can see, the last digit of 7744 is 4. So from Table 1, we get two possible sets of answers for **b**, i.e. 2 and 8.

Multiply a = 8 by a + 1 = 9

$8 \times 9 = 72 < 88$ (First pair of group)

Hence b = 8 (Higher of two possible values)

Therefore, $\sqrt{7744} = 88$

Now, the whole process laid out above can be done with a minor change in it. Let's do the three examples given above with a new method.

Method 3

- Find the value of a and b as done in Method 2.
- Put b = 5 and find the square = $(a\,5)^2$
- If original number > $(a\,5)^2$ then take the smaller value of b
- If original number < $(a\,5)^2$ then take the larger value of b

Example: Find the square root of 7744.
Solution: We have two groups here—77 and 44.

$8^2 < 77 < 9^2$

Hence, a = 8

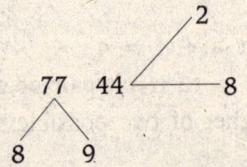

As you can see, the last digit of 7744 is 4. So from Table 1, we get two possible sets of answers for **b**, i.e. 2 and 8. Hence, $(7744)^{1/2} = 82$ or 88

Now find $(85)^2 = 7225^*$

Since, $7744 > 7225$

Hence b is maximum or b = 8

$(7744)^{1/2} = 88$

Example: Find the square root of 1296.
Solution: We have two groups here—12 and 96. Ignore the last group 96 and find the value of a for the first group.

$3^2 < 12 < 4^2$

Hence, a = 3

As you can see, the last digit of 1296, that is the unit digit of the second group 96, is 6. So from Table 1, we get two possible sets of answers for b, i.e. 4 and 6.

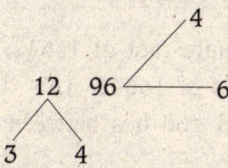

The two possible answers of $\sqrt{1296}$ are 34 or 36.

$(35)^2 = 1225$

$1296 > 1225$

Hence, $\sqrt{1296} = 36$

(Maximum value of b out of 4 and 6)

Example: Find the square root of 4489.
Solution: The first pair 44 lies between the square of 6 and 7.

a = Minimum value = 6

---

*The method of squaring a number ending with 5 has been discussed in the chapter on Squares.

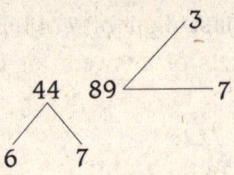

Last digit of 4489 is 9, so b can have 3 or 7.
  $\sqrt{4489}$ = 63 or 67
  Now, $(65)^2$ = 4225 < 4489
  Hence, b = 7 (Maximum value)

Finding the Square Root of Five-digit Numbers

In this case, the method is the same. Take the first three digits (from the left) to determine 'a', and 'b' can be determined with the method discussed above.

Let's look at a few examples.

Example: Find the square root of 16641.
Solution: Make 2 pairs of 16641. This time, the first pair is a three-digit number and 166 lies between the square of 12 and 13. Hence a = 12.

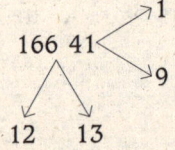

Moreover, unit digit is 1, so b = 1 or 9.
  $\sqrt{16641}$ = 121 or 129
  Now, $(125)^2$ = 15625
  16641 > 15625
  Hence, b = 9 (Maximum value)
  $\sqrt{16641}$ = 129

Example: Find the square root of 61504.
Solution: Make 2 pairs of 61504. This time, the first pair is a

three-digit number and 615 lies between the square of 24 and 25. Hence a = 24.

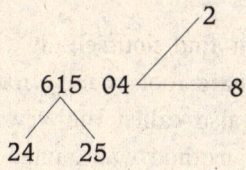

Moreover, unit digit is 4, so b = 2 or 8.
 √61504 = 242 or 248
 Now, (245)² = 60025
 61504 > 60025
 Hence, b = 8 (Maximum value)
 √61504 = 248

Example: Find the square root of 29929.
Solution: Make 2 pairs of 29929. The first pair 299 lies between the square of 17 and 18. Hence a = 17.

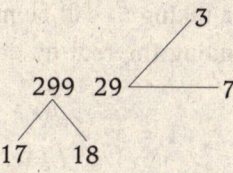

Moreover, unit digit is 9, so b = 3 or 7.
 √29929 = 173 or 177
 Now, (175)² = 30625
 29929 < 30625
 Hence, b = 3 (Minimum value)
 √29929 = 173

I hope you have enjoyed this beautiful method of finding the square root. However, before you attempt the above mentioned methods you should master those that are used to find squares!

Finding Square Roots Using the Newton's Method or the Newton-Raphson Method

Now, many times you find yourself in a situation where you need to guess the square root or cube root of a number. Here, the Newton Method, also called the Newton–Raphson Method, comes in handy. This method was named after mathematicians Isaac Newton and Joseph Raphson. It is a method for finding successively better approximations of any roots, whether it is a square root or cube root. This method works well if you can guess the first approximate value that is closer to the value whose square root or cube root needs to be extracted.

Let's suppose that we have to find the square root of 35. So the nearest value for the first approximate may be 5 or 6 as the square of 5 is 25 whereas the square of 6 is 36. Since 35 is very close to 36, 6 can be considered to be a better guess. Though you will get the answer in both the cases, the number of iterations, if you are taking 5, will be more as compared to 6.

The formula for finding the root by the Newton's Method is:

$$x_{n+1} = x_n - \frac{f(x_n)}{f'(x_n)}$$

Let's begin with a simple example.

Example: Find the square root of 2.
Solution: Let $f(x) = x^2 - 2$
So $f'(x) = 2x$
Now make a table.

X	$f(x) = x^2 - 2$	$f'(x) = 2x$	X − f(x) / f'(x)
1	−1	2	$1 + 1/2 = 3/2 = 1.5$
1.5	$1/4$	3	17/12 = 1.41
1.41	1/144	17/6	577/408 = 1.414

Since the answer in the last two iterations is closer, hence we can assume $\sqrt{2} = 1.41$

Example: Find the square root of 10.
Solution: The nearest square of 10 is 3, so we shall begin with 3.

X	$f(x) = x^2 - 10$	$f'(x) = 2x$	$X - f(x)/f'(x)$
3	-1	6	$3 - 1/6 = 3.1666$
3.16			3.162
3.162			3.162

Note that we can use the Newton's Method in finding cube roots too.

Example: Find the cube root of 406.
Solution: 406 is closer to 343 and the cube root of 343 is 7. So we will have to take $x = 7$ as the first iteration.

X	$f(x) = x^3 - 406$	$f'(x) = 3x^2$	$X - f(x)/f'(x)$
7	-63	147	$7 + 0.42 = 7.42$
7.4	-0.77	164.28	7.39

Hence, the cube root of 406 is approximately equal to 7.4.

As you can see, it involves a lot of calculation. So let's improvise the whole process with a small change.

Working Rule:

a) Guess a perfect square root as close as possible to your number.
b) Divide the number by the first guess.
c) Calculate the average of the result of step 2 and the root.
d) Repeat step 2 and 3 until you have a number that is accurate enough for you.

Example: Find the square root of 695.
Solution: We know the square root of 25 is 625; so our first guess is 25. You may, however, take $26^2 = 676$

Let us first guess the root as 25.

1. Divide the number by first guess = 695 ÷ 25 = 27.8
2. Take average of first guess and the previous result obtained = (25 + 27.8) ÷ 2 = 26.4
   Then, repeat the process.
3. 695 ÷ 26.4 = 26.3257
4. (26.4 + 26.3257) ÷ 2 = 26.36285
5. 695 ÷ 26.36285 = 26.3628553

The results in the last two steps are almost similar; so the square root of 695 = 26.36

Example: Find the square root of 2603.
Solution: The nearest square that you can guess is 2500 whose square root is 50.

Now follow the instructions:

1. Divide the number by first guess = 2603 ÷ 50 = 52.06
2. Take the average of the first guess and the previous result = (50 + 52.06)/2 = 51.03
3. Divide the number by the second guess = 2603 ÷ 51.03 = 51.009
4. Take the average = (51.03 + 51.009)/2 = 51.01

Hence, √2603 = 51.01

This method of estimating roots seems faster than the Newton's Method. Isn't it?

Practice Problems

Find the square root of the following by guessing:

a) 37   b) 47   c) 78   d) 111
e) 679   f) 1201   g) 2091   h) 4508

Find the square root of the following by applying the methods learnt in the chapter:

a) 2116
b) 4225
c) 6889
d) 59049
e) 125316
f) 9801
g) 2304
h) 37
i) 164
j) 17
k) 43
l) 198

Cube

Cubing a number means multiplying a number with itself three times. In general, one needs to cube a number to find the volume of spheres and hemispheres, to calculate compound interests (when time is of 3 years) and moreover, in competitive examinations, when questions relate to basic algebraic identities.

Let's take this example.

$$\frac{0.8 \times 0.8 \times 0.8 - 0.6 \times 0.6 \times 0.6}{0.8 \times 0.8 + 0.8 \times 0.6 + 0.6 \times 0.6}$$

Solution: You can solve such a question by solving the numerator and denominator separately and for that, you need to know the concepts of cube and square. But if you are good at algebra, then you can solve it very quickly.

Let $a = 0.8$ and $b = 0.6$

$$\text{Expression} = \frac{a^3 - b^3}{a^2 - ab + b^2}$$
$$= \frac{(a-b)(a^2 + ab + b^2)}{a^2 + ab + b^2}$$
$$= a - b$$
$$= 0.8 - 0.6 = 0.2$$

Take another example.

$$\frac{4.2 \times 4.2 \times 4.2 + 1.8 \times 1.8 \times 1.8 + 2.1 \times 2.1 \times 2.1 - 3 \times 4.2 \times 1.8 \times 2.1}{4.2 \times 4.2 + 1.8 \times 1.8 + 2.1 \times 2.1 - 4.8 \times 1.8 - 1.8 \times 2.1 - 2.1 \times 4.8}$$

Solution: If $x = 4.2$, $y = 1.8$, $z = 2.1$ then the above expression can be written as:

$$\frac{x^3 + y^3 + z^3 - 3xyz}{(x^2 + y^2 + z^2 - zy - yz - zx)}$$

$$= \frac{(x + y + z)(x^2 + y^2 + z^2 - zy - yz - zx)}{(x^2 + y^2 + z^2 - zy - yz - zx)}$$

$$= x + y + z$$
$$= 4.2 + 1.8 + 2.1$$
$$= 8.1$$

Example:

$$\frac{343 \times 343 \times 343 - 113 \times 113 \times 113}{343 \times 343 + 343 \times 113 + 113 \times 113}$$

Solution: We know that
$$a^3 - b^3 = (a - b)(a^2 + ab + b^2)$$

If $a = 343$ and $b = 113$, then the above expression will become

$$\frac{a^3 - b^3}{a^2 + ab + b^2}$$

So $\dfrac{a^3 - b^3}{a^2 + ab + b^2} = \dfrac{(a - b)(a^2 + ab + b^2)}{a^2 + ab + b^2} = a - b = 343 - 113 = 230$

Had you not used the algebraic formula, you would have solved these equations by finding the cube of numbers and putting the values in the given expression.

Now, let's explore the method of cubing.

In all the methods that I will discuss here, I shall try to use concepts of algebra.

Case 1: Finding the Cube Using the Formula
$a^3 = (a - 1) \times a \times (a + 1) + a$

I love this formula! Calculating the cube of any number using this formula is very easy. Moreover, if you know multiplying three numbers in a single go using the Vedic sutra, you can sum up the whole process in a few seconds.[*] Let's begin with a simple example: Find the cube of 11.

Solution: Here, a = 11
$$(11)^3 = 10 \times 11 \times 12 + 11$$
$$= 1320 + 11$$
$$= 1331$$

(Here, multiply 12 and 10 first and then apply the method of multiplying with 11.)

Example: Find the cube of 12.
Solution: Here, a = 12
$$(12)^3 = 11 \times 12 \times 13 + 12$$
$$= 1716 + 12$$
$$= 1728$$

(Use the method of multiplying two numbers close to base 10 and multiply 12 and 13. Use property of 11.)

Example: Find the cube of 15.
Solution: Here, a = 15
$$(15)^3 = 14 \times 15 \times 16 + 15$$
$$= 3360 + 15$$
$$= 3375$$

(Unit digits of 14 and 16 sum up to 10 [4 + 6 = 10]; multiply these numbers and then multiply the product with 15.)

Example: Find the cube of 25.
Solution: Here, a = 25
$$(25)^3 = 24 \times 25 \times 26 + 25$$
$$= 15600 + 25$$
$$= 15625$$

(24 and 26 can be multiplied easily as their unit digit sum is 10. Later, use the method of multiplying with 25.)

---

[*] In my book *The Essentials of Vedic Mathematics*, I have given a detailed explanation of multiplying three or four numbers at a go.

Example: Find the cube of 99.
Solution: Here, a = 99

$(99)^3$ = 98 × 99 × 100 + 99
     = 970200 + 99
     = 970299

Case 2: Using the Formula $(a + b)^3 = a^3 + 3a^2b + 3ab^2 + b^3$

This one is the basic formula you have studied in your upper-primary class. Let's explore it.

Example: Find the cube of 12.
Solution: Here, a = 1 and b = 2

- First find $b^3 = 8$
- $3ab^2 = 3 \times 1 \times 2^2 = 12$
- $3a^2b = 3 \times 1^2 \times 2 = 6$
- $a^3 = 1^3 = 1$
  $(ab)^3 = a^3 / 3a^2b / 3ab^2 / b^3$
     = 1 / 6 / 12 / 8
     = 1 / 6 + 1 / 2 / 8
     = 1728

Example: Find the cube of 25.
Solution: Here, a = 2 and b = 5

- First find $b^3 = 125$
- $3ab^2 = 3 \times 2 \times 5^2 = 150$
- $3a^2b = 3 \times 2^2 \times 5 = 60$
- $a^3 = 2^3 = 8$
  $(ab)^3 = a^3 / 3a^2b / 3ab^2 / b^3$
     = 8 / 60 / 150 / 125
     = 8 / 60 / 150 + 12 / 5
     = 8 / 60 / 162 / 5
     = 8 / 60 + 16 = 76 / 2 / 5
     = 8 + 7 / 6 / 2 / 5
     = 15625

Example: Find the cube of 47.
Solution: Here, a = 4 and b = 7

- First find $b^3 = 343$
- $3ab^2 = 3 \times 4 \times 7^2 = 588$
- $3a^2b = 3 \times 4^2 \times 7 = 336$
- $a^3 = 4^3 = 64$

$(ab)^3 = a^3 / 3a^2b / 3ab^2 / b^3$
$ = 64 / 336 / 588 / 343$
$ = 64 / 336 / 588 + 34 = 622 / 3$
$ = 64 / 336 + 62 = 398 / 2 / 3$
$ = 64 + 39 / 8 / 2 / 3$
$ = 103823$

Case 3: Using $(a + b)^3 = a^3 + 3a^2b + 3ab^2 + b^3$ in a Different Way

Let's first expand $(1 + r)^3$:
$(1 + r)^3 = 1 + 3r + 3r^2 + r^3$
This can be written differently in the following way.

$(1 + r)^3$	1	r	r^2	r^3
+		2r	$2r^2$	
	1	3r	$3r^2$	r^3

The table shows that if we take b/a = r and make a four-columned table leaving the first to contain only the problem at hand (so we treat the second column as the first), we will then have to double the entries in the second and third columns in the table and add the results; thereby we get the exact formula of $(a + b)^3$.

Let's see the working with an example.

Example: Find the cube of 11.
Solution: Here a = 1 and b = 1. Hence r = b /a = 1/ 1 = 1.

Fill the first column like it's shown above. Double the entries of the second and third columns and add the results.

$(11)^3$	1	1	1	1
+		2	2	
	1	3	3	1

Example: Find the cube of 12.
Solution: Here a = 1 and b = 2. Hence r = b / a = 2 / 1 = 2.

Fill the first column as explained above. Double the entries of the second and third columns and add the results up.

$(12)^3$	1	2	4	8
+		4	8	
	1	6	12	8
	1	6 +1 = 7	2	8

(**Note:** In each column, except the leftmost column, there should be a single digit. If you have more than 1 digit in any of the columns, transfer the leftmost digit(s) to the previous column.)

Here, in the third column, the sum of 4 + 8 = 12. Put only 2 in the third column and transfer the leftmost digit 1 to the second column and add it with the number already written in the second column. Therefore, in the second column we have 6 + 1 = 7.

Example: Find the cube of 15.
Solution: Here a = 1 and b = 5. Hence r = b / a = 5 / 1 = 5.

Fill the first column as explained above. Double the entries of the second and third columns and add the results.

$(15)^3$	1	5	25	125
+ .		10	50	
	1	15	75	125
	1	15	75 + 12 = 87	5
	1	15 + 8 = 23	7	5
	1 + 2 = 3	3	7	5

Here, in the fourth column, we have 125. Keep 5 in the fourth column and transfer 12 to the third column. Third column

Cube 223

= 75 + 12 = 87. Keep 7 in the third column and transfer 8 of 87 to the second and add it to 15 from the original entries in the second column, making the total value of the second column 15 + 8 = 23. Keep 3 of 23 in the second column and transfer 2 to the first column, making the total sum of the first column 1 + 2 = 3.

Example: Find the cube of 25.

Solution: Here $a = 2$ and $b = 5$. Hence $r = b/a = 5/2$

Fill the first column as explained above. Double the entries of the second and third columns and add the results.

$(25)^3$	$2^3 = 8$	$8 \times 5/2 = 20$	$20 \times 5/2 = 50$	125
+		40	100	
	8	60	150	125
	8	60	150 + 12 = 162	5
	8	60 + 16 = 76	2	5
	8 + 7 = 15	6	2	5

Here, in the fourth column, we have 125. Keep 5 in the fourth column and transfer 12 to the third column. Third column = 150 + 12 = 162. Keep 2 in the second column and transfer 16 of 162 to the second and add it to 60 from the original entries in the second column making the total value of the second column 60 + 16 = 76. Keep 6 of 76 in the second column and transfer 7 to the first column, making the total sum of the first column 8 + 7 = 15.

Case 4: Using $(a + b)^3 = a^3 + 3a^2b + 3ab^2 + b^3$ in a Different Way

We know, $(a + b)^3 = a^3 + 3a^2b + 3ab^2 + b^3$
$$= a^2(a + 3b) + 3b^2 \times a + b^3$$
$$= a^2(a + b + 2b) + 3b^2 \times a + b^3$$

This method will work better if the number is near base 10, 100, 1000.

How does it work?

First, take the deviation of the number to be cubed from its base. The base should be the multiple of 10. If the base is 10, 100, 1000, etc., then sub-base = 1; on the other hand, if the base = 20, then the sub-base = 2, because 20 = 2 × 10.

The whole cubing process then involves 3 steps.

- $a^2 (a + b + 2b)$ = (Number to be cubed + 2 × deviation from the base) × (sub-base)^2
- $3b^2 \times a$ = {3 × (deviation)^2} × sub-base
- b^3 = (Deviation)^3

Example: Find the cube of 12.
Solution: Base = 10

Hence, deviation = 12 − 10 = 2

Step 1 = $a^2 (a + b + 2b)$ = (Number to be cubed + 2 × deviation from the base) × (sub-base)^2
= (12 + 2 × 2) × 1^2
= 16

Step 2 = $3b^2 \times a$ = {3 × (deviation)^2} × sub-base
= (3 × 2^2) × 1
= 12

Step 3 = b^3 = (Deviation)^3
= 2^3 = 8

Hence, $(12)^3$ = 16 / 12/ 8
= 1728

(Keep only one digit in each part and transfer the excess to the previous part.)

Example: Find the cube of 15.
Solution: Base = 10

Hence, deviation = 15 − 10 = 5

Step 1 = $a^2 (a + b + 2b)$ = (Number to be cubed + 2 × deviation from the base) × (sub-base)^2

$$= (15 + 2 \times 5) \times 1^2$$
$$= 25$$

Step 2 $= 3b^2 \times a = \{3 \times (\text{deviation})^2\} \times \text{sub-base}$
$$= (3 \times 5^2) \times 1$$
$$= 75$$

Step 3 $= b^3 = (\text{Deviation})^3$
$$= 5^3 = 125$$

Hence, $(15)^3 = 25/75 / 125$
$$= 25/75 + 12 = 87 / 5$$
$$= 25 + 8 / 7 / 5$$
$$= 3375$$

Example: Find the cube of 43.

Solution: Base = 40

Hence, deviation = 43 – 40 = 3, Sub-base = 4 (40 = 4 × 10)

Step 1 $= a^2 (a + b + 2b) = $ (Number to be cubed + 2 × deviation from the base) × (sub-base)^2
$$= (43 + 2 \times 3) \times 4^2$$
$$= 49 \times 16 = 784$$

Step 2 $= 3b^2 \times a = \{3 \times (\text{deviation})^2\} \times \text{sub-base}$
$$= (3 \times 3^2) \times 4$$
$$= 108$$

Step 3 $= b^3 = (\text{Deviation})^3$
$$= 3^3 = 27$$

Hence, $(43)^3 = 784 / 108/ 27$
$$= 784 / 108 + 2 = 110 / 7$$
$$= 784 + 11 / 0 / 7$$
$$= 79507$$

Example: Find the cube of 92.

Solution: Base = 90, Hence, deviation = 92 – 90 = 2,
Sub-base = 9 (90 = 9 × 10)

Step 1 $= a^2 (a + b + 2b) = $ (Number to be cubed + 2 × deviation from the base) × (sub-base)^2

$$= (92 + 2 \times 2) \times 9^2$$
$$= 96 \times 81 = 7776$$

Step 2 = $3b^2 \times a$ = {$3 \times (\text{deviation})^2$} × sub-base
$$= (3 \times 2^2) \times 9$$
$$= 108$$

Step 3 = b^3 = $(\text{Deviation})^3$
$$= 2^3 = 8$$

Hence, $(92)^3$ = 7776/ 108/ 8
$$= 7776 + 10 / 8 / 8$$
$$= 7786 / 8 / 8$$

Example: Find the cube of 98.

Solution: 98 is nearer to the base 100.

Deviation = 98 − 100 = −2

Hence $(98)^3$ = $\underbrace{98 + 2 \times (-2)}_{\text{1st term}} \mid \underbrace{3 \times (-2)^2}_{\text{2nd term}} \mid \underbrace{(-2)^3}_{\text{3rd term}}$

$$= 94 \mid 12 \mid -8 = 94 \mid 11 \mid 100 - 8$$
$$= 94 \mid 11 \mid 92$$

Hence $(98)^3$ = 941192

Example: Find the cube of 104.

Solution: Working base = 100

Deviation = 104 − 100 = 4

$(104)3 = 104 + 4 \times 2 \mid 3 \times 4^2 \mid 4^3$
$$= 112 \mid 48 \mid 64$$
$$= 1124864$$

(Since the base = 100, there should be 2 digits in each digit separator.)

The four methods discussed in this chapter will undoubtedly help you to find cubes of numbers. But I would suggest to you to also have imprinted in your minds the cube of the first 10 numbers so that while extracting the cube root of any number, you can directly access that knowledge base and get the cube root in seconds.

Now, let's learn something different.

$1^3 + 2^3 = (1 + 2)^2 = 3^2$
$1^3 + 2^3 + 3^3 = (1 + 2 + 3)^2 = 6^2$
$1^3 + 2^3 + 3^3 + 4^3 = (1 + 2 + 3 + 4)^2 = 10^2$
$1^3 + 2^3 + 3^3 + 4^3 + 5^3 = (1 + 2 + 3 + 4 + 5)^2 = 15^2$
$1^3 + 2^3 + 3^3 + 4^3 + 5^3 + 6^3 = (1 + 2 + 3 + 4 + 5 + 6)^2 = 21^2$

If you observe the right-hand side minutely, you will find the triangular numbers 1, 3, 6, 10, 15, 21.

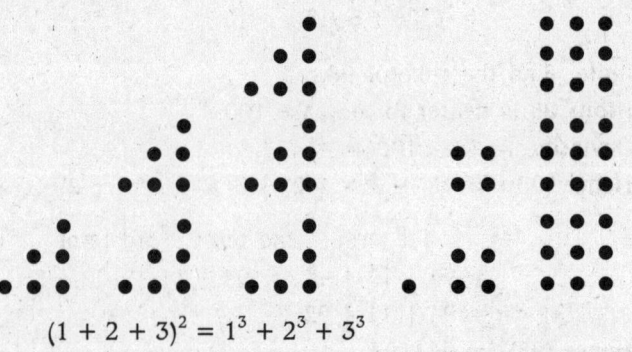

$$(1 + 2 + 3)^2 = 1^3 + 2^3 + 3^3$$

Hence,
$$(1 + 2 + 3 + \ldots + n)^2 = \frac{n^2(n+1)^2}{4} = 1^3 + 2^3 + \ldots + n^3$$

Let's try—
$$1^3 + 2^3 + \ldots 10^3 = \frac{10^2 \times 11^2}{4}$$
$$= 3025 = (1 + 2 + \ldots + 10)^2 = 55^2$$

If $C_1, C_2, C_3,$ etc. are the first, second, third cubic numbers, then they exhibit a unique property:

$$C_1 = (1)^2 = (T_1)^2$$
$$C_1 + C_2 = 1 + 8 = 9 = (T_2)^2$$
$$C_1 + C_2 + C_3 = 1 + 8 + 27 = 36 = (T_3)^2$$
$$C_1 + C_2 + C_3 + C_4 = 1 + 8 + 27 + 64 = 100 = (T_4)^2$$

A square number shows different unique properties. It can be written as the sum of cubic numbers and triangular numbers.

The interesting thing that you should know is that it was Pythagoras, the Greek mathematician, who had introduced square numbers, triangular numbers, cubic numbers and many more. Keep enjoying!

Practice Problems

Find the cube of each of the following:

a) 13   b) 19   c) 25   d) 36   e) 46   f) 54
g) 69   h) 87   i) 104 j) 113 k) 208 l) 315

Cube Roots

Let's begin with an example. This question was asked some years back in a competitive examination.

Find the value of $\sqrt[3]{681472} - \sqrt{6084} + \sqrt[3]{117649}$

In order to solve the given expression, you need to know the speedy method to calculate cube roots since in competitive examinations, time management is one of the few ways to success.

So, if $y = x^3$

Then $x = (y)^{1/3}$

Let me first begin with those numbers which are not a perfect cube root.

Cube Roots of Numbers that are Not Perfect Cubes

First off, here is a ready reference for finding cube roots of any number.

Table 1

$\sqrt[3]{1} = 1$	$\sqrt[3]{8} = 2$
$\sqrt[3]{27} = 3$	$\sqrt[3]{64} = 4$
$\sqrt[3]{125} = 5$	$\sqrt[3]{216} = 6$
$\sqrt[3]{343} = 7$	$\sqrt[3]{512} = 8$
$\sqrt[3]{729} = 9$	$\sqrt[3]{1000} = 10$

If a number is not a perfect cube then its approximate answer can be obtained with the help of the formula given below.

> Cube root of a number = Nearest cube root $\pm \dfrac{\text{Difference from the original number}}{3 \times (\text{Nearest cube root})^2}$

Example: Find the cube root of 29.

Solution: 27 is closest to 29 and cube root of 27 is 3.

Difference, $29 - 27 = 2$

Now apply the above formula.

$(29)^{1/3} = 3 + \dfrac{2}{3 \times 3^2}$

$= 3 + 2/27$

$= 3 + 0.74$

$= 3.74$

Example: Find the cube root of 150.

Solution: 125 is closest to 150 and cube root of 125 is 5.

Difference, $150 - 125 = 25$

Now apply the above formula.

$(150)^{1/3} = 5 + \dfrac{25}{3 \times 5^2}$

$= 5 + 1/3$

$= 5 + 0.33$

$= 5.33$

Example: Find the cube root of 214.

Solution: 216 is closest to 214 and the cube root of 216 is 6.

Difference, $214 - 216 = -2$

Now apply the above formula.

$(214)^{1/3} = 6 - \dfrac{2}{3 \times 6^2}$

$= 6 - 1/54$

$= 6 - 0.018$

$= 5.982$

Example: Find the cube root of 541.

Solution: 512 is closest to 541 and the cube root of 512 is 8.

Difference, $541 - 512 = 29$

Cube Roots

Now apply the above formula.

$$(541)^{1/3} = 8 + \frac{29}{3 \times 8^2}$$
$$= 8 + {}^{29}/_{192}$$
$$= 8 + 0.15$$
$$= 8.15$$

Now let's move to find the cube root of those numbers which are perfect cubes.

Cube Roots of Numbers that are Perfect Cubes

Cube root of numbers 4–6 digits long uses the formula:
$(a + b)^3 = a^3 + 3a^2b + 3ab^2 + b^3$

Rule:

- First make a group of 3 digits each. From the cube root table find the nearest cube root for the first pair. Let it be 'a'.
- Subtract a^3 and put the remainder, if any, before the next digit of the number whose cube root you want to find.
- Equate the next dividend to $3a^2b$ and find 'b'. (Take only the whole part of the answer; for instance, if the result is 3.05, then only 3 will be taken.)
- Subtract $3a^2b$ from the dividend now. Place the remainder before the next digit.
- Now subtract $3ab^2$ from the new dividend and place the remainder before the next digit.
- Finally subtract b^3 to get the remainder zero (0).

Let's have this applied to some examples.

Example: Find the cube root of 1728.
Solution: First make the pair

$\overline{1}\ \overline{728}$

The first pair is 1 and second pair is 728.
Since 1 lies between the cubes of 1 and 2, a = 1 (lower number)

Subtract a^3 from 1.

$1 - 1^3 = 0$

Put down the next number 7 from the dividend.

New dividend = 7

Equate the new dividend to $3a^2b$; $3 \times 1^2 \times b = 7$

So $b = 2$

Subtract to $3a^2b = 6$ from the new dividend 7

Remainder = $7 - 6 = 1$

Put down the next number (2) from the dividend next to 1.

New dividend = 12

Subtract $3ab^2 = 3 \times 1 \times 2^2 = 12$ from the new dividend = $12 - 12 = 0$

Carry down the next digit (8) from the dividend

```
   1̄ 7̄2̄8̄
  -1  ─────────────── a³
  ──
   07
   -6  ─────────────── 3a²b
   ──
   12
  -12  ─────────────── 3ab²
   ──
    8
   -8  ─────────────── b³
   ──
    x
```

New dividend = $8 = 2^3$

Subtract b^3 from the dividend = $8 - 8 = 0$

Hence, Cube Root of $1728 = 12$

Example: Find the cube root of 2197.

Solution: First make the pair.

$\overline{2}\ \overline{197}$

The first pair is 2 and second pair is 197.

Since 2 lies between the cubes of 1 and 2 so $a = 1$ (lower number)

Subtract a^3 from 1.

$2 - 1^3 = 1$

Put down the next number 1 from the dividend.

New dividend = 11
Equate the new dividend to $3a^2b$
$3 \times 1^2 \times b = 11$
So b = 3
Subtract to $3a^2b = 9$ from the new dividend 11
Remainder = 11 − 9 = 2
Put down the next number (9) from the dividend next to 2
New dividend = 29
Subtract $3ab^2 = 3 \times 1 \times 3^2 = 27$ from the new dividend
= 29 − 27 = 2
Carry down the next digit (7) from the dividend
New dividend = 27 = 3^3
Subtract b^3 from the dividend = 27 − 27 = 0
Hence, Cube Root of 2197 = 13

Example: Find the cube root of 13824.
Solution: First make the pair
$\overline{13}\ \overline{824}$

Since 13 lies between the cubes of 2 and 3 so a = 2 (lower number)

Subtract a^3 from 13
$13 - 2^3 = 5$
Put down the next number 8 from the dividend.
New dividend = 58
Equate the new dividend to $3a^2b$
$3 \times 2^2 \times b = 58$
So b = 4
Subtract $3a^2b = 48$ from the new dividend 58
Remainder = 58 − 48 = 10
Put down the next number (2) from the dividend next to 10
New dividend = 102
Subtract $3ab^2 = 3 \times 2 \times 4^2 = 96$ from the new dividend
= 102 − 96 = 6
Carry down the next digit (4) from the dividend
New dividend = 64 = 4^3

Subtract b^3 from the dividend = 64 − 64 = 0

```
 13 824
 −8  ─────────────────  a³
 ──
 58
 −48  ─────────────────  3a²b
 ───
 102
 −96  ─────────────────  3ab²
 ───
  64
 −64  ─────────────────  b³
 ───
   x
```

Hence, Cube Root of 13824 = 24

Example: Find the cube root of 884736.

Solution: First make the pair.

$\overline{884}\ \overline{736}$

Since 884 lies between the cubes of 9 and 10 so a = 9 (lower number)

Subtract a^3 from 884, i.e.

$884 − 9^3 = 155$

Put down the next number 7 from the dividend.

New dividend = 1557

Equate the new dividend to $3a^2b$

$3 \times 9^2 \times b = 1557$

So b = 6

Subtract $3a^2b = 1458$ from the new dividend 1557

Remainder = 1557 − 1458 = 99

Put down the next number (3) from the dividend next to 99

New dividend = 993

Subtract $3ab^2 = 3 \times 9 \times 6^2 = 972$ from the new dividend = 993 − 972 = 216

Carry down the next digit (6) from the dividend

New dividend = 216 = 6^3

Subtract b^3 from the dividend = 216 − 216 = 0

$$
\begin{array}{rl}
884736 & \\
-729 & \text{———} \quad a^3 \\
\hline
1557 & \\
-1458 & \text{———} \quad 3a^2b \\
\hline
993 & \\
-972 & \text{———} \quad 3ab^2 \\
\hline
216 & \\
-216 & \text{———} \quad b^3 \\
\hline
\text{xx} &
\end{array}
$$

Hence, Cube Root of 884736 = 96

However sharp, the method discussed above is not feasible as it takes time to arrive at the answer. Let's take a method through which it is easier to find the cube roots of numbers less than 7 digits.

Important Points to Remember

- The cubes of the first nine natural numbers are as follows:

M	1	2	3	4	5	6	7	8	9
M^3	1	8	27	64	125	216	343	512	729

- The number of digits in a cube root is the same as the number of three-digit groups in the original cube, including a single-digit or a double-digit group, if there exists any.
- 1, 4, 5, 6, 9 and 0 repeat themselves in the cube-ending.
- 2, 3, 7 and 8 have an interplay of complements from 10. In other words, the cubes of 2, 3, 7 and 8 end with 8, 7, 3 and 2 respectively.

Table 2

Unit Digit of a Cube Root	Unit Digit of a Cube
1	1
8	2
7	3
4	4

Contd. on p. 240

Contd. from p. 239

5	5
6	6
3	7
2	8
9	9

- The first digit (from the left) of the cube root will always be obvious from the first group in the cube.
- Thus, the number of digits, the first digit and the last digit of the cube root of an exact cube is the data with which we start when we begin the work of extracting the cube root of an exact cube.
- Most importantly, the digit sum of any number[*] will always be 1, 8 or 9(0).

  Example: The digit sum of 314432 is $3 + 1 + 4 + 4 + 3 + 2 = 8$; therefore, 314432 is the perfect cube root.

Case 1: Finding the Cube Root of a Number Having Less Than 7 Digits

The cube root of numbers having 4 digits, 5 digits and 6 digits will have 2 digits in its cube root. The unit digit can be obtained from Table 2 and the tens' digit can be obtained from Table 1. It will simply take less than 5 seconds to guess the exact cube root for those numbers. Follow the steps and get the answer instantly.

- Look at the digits and determine the digit in the units' place using Table 2.
- Place a bar from right to left, grouping three numbers at a time.

For example:

$$\overline{3}\ \overline{431}\ \overline{331}$$

---

[*]The digit sum of a number can be obtained by using the Casting Out Nines method described in this book in detail.

- If no digit is left (as in the case of 343), the digit obtained in step 1, is the required cube root.
- If digits are left, find the largest single-digit number whose cube root is less than or equal to this leftover number.

Now, let us take a few examples.

Example: Find the cube root of 389017.
Solution:
1. Place the bar over the number from left to right, leaving two digits at a time.

$$\overline{389}\ \overline{017}$$

2. Since the bar is placed on two numbers, the cube root will contain only two digits.
3. Since the unit digit of this number is 7, on comparing the table one can see that the unit digit of this number is 3.

$$\overline{389}\quad \overline{017}$$
$$7^3 \quad 8^3 \qquad 3$$

4. For the ten's digit, take the least of
$7^3 < 389 < 8^3$
Hence the tens' digit = 7
$\Rightarrow \sqrt[3]{389017} = 73$

Example: Find the cube root of 328509.
Solution:
1. Place the bar over the number from left to right, leaving two digits at a time.

$$\overline{328}\ \overline{509}$$

2. Since the bar is placed on two numbers, the cube root will contain only two digits.

3. Since the unit digit of the number 328509 is 9, on comparing this with Table 2, we see that the unit digit of this number will be 9.

$$\overline{328} \quad \overline{509}$$
$$\diagup\diagdown \quad \mid$$
$$216 \quad 343 \quad 9$$

4. For the tens' digit, take the least of
$6^3 < 389 < 7^3$
Hence the tens' digit = 6
$\Rightarrow \sqrt[3]{328509} = 69$

Example: Find the cube root of 658503.
Solution:
1. Place the bar over the number from left to right, leaving two digits at a time.
2. Since the bar is placed on two numbers, the cube root will contain only two digits.
3. Since the unit digit of this number is 3, upon comparing, the table shows that the unit digit of this number is 7.

$$\overline{658} \quad \overline{503}$$
$$\diagup\diagdown \quad \mid$$
$$512 \quad 729 \quad 7$$

4. For the tens' digit, take the least of
$8^3 < 658 < 9^3$
Hence the tens' digit = 8
$\Rightarrow \sqrt[3]{658503} = 87$

Isn't the method discussed above easy?

The cube root of any number less than 7 digits can be extracted with the help of the two tables given in this chapter: Table 1 could be used to find the tens' digit and Table 2 to find the unit digit of the cube root. Keep practising!

Practice Problems

Find the cube root of the following numbers:

a) 6892 b) 636056 c) 314432 d) 8365427
e) 1061208 f) 8489664 g) 143055667 h) 9800344
i) 300763 j) 195112 k) 328509 l) 551368